Modules 1 – 6

FOUNDATION

# MODULAR SCIENCE

## for Edexcel





Graham Booth
Bob McDuell

Heinemann

Heinemann Educational Publishers
Halley Court, Jordan Hill, Oxford OX2 8EJ
a division of Reed Educational & Professional Publishing Ltd
Heinemann is a registered trademark of Reed Educational & Professional Publishing Ltd

OXFORD MELBOURNE AUCKLAND
JOHANNESBURG BLANTYRE GABORONE
IBADAN PORTSMOUTH NH (USA) CHICAGO



First published 2002

ISBN 0 435 584669

05 04 03 02
10 9 8 7 6 5 4 3 2 1

Edited by Helen Barham, Helen Roberts and Margaret Shepherd

Designed and typeset by Hardlines, Charlbury

Illustrated by Hardlines, Charlbury

Printed and bound in Spain by Edelvives

**Acknowledgements**
The authors and publishers would like to thank the following for permission to use photographs:

Cover photos: Scan of red blood cells courtesy of Science Photo Library/Andrew Syred; Rock salt crystals courtesy of Science Photo Library/Charles D. Winters; Wind turbines courtesy of Science Photo Library/Chris Knapton.

***T = top B = bottom R = right L = left M = middle***

**1** Photodisc; **2** Art Directors & Trip; **6** Science Photo Library; **8** ***(both)*** Science Photo Library; **10** Science Photo Library; **12** ***(both)*** Science Photo Library; **13** Panos Pictures; **16** ***(all)*** Photodisc; **20** Science Photo Library; **23** Panos Pictures; **24** Science Photo Library; **26** Oxford Scientific Films/Max Gibbs; **29** Science Photo Library; **33** Photodisc; **34** ***T*** Oxford Scientific Films/Colin Monteath, ***B*** Image Bank/Vikki Hart; **35** Science Photo Library/CNRI; **38** ***T*** Photodisc, ***B*** Sally & Richard Greenhill/Sally Greenhill; **42** Photodisc; **44** HGP; **45** Photodisc; **46** ***L*** Ace/Bo Cederwall, ***R*** Rex Features, **48** Oxford Scientific Films/Tony Tilford; **49** Empics/Tony Marshall; **50** ***T*** Holt Studios/Nigel Cattlin, ***B*** Holt Studios/Nigel Cattlin; **51** Photodisc; **52** ***T*** BBC Wild/Mark Payne-Gill, ***BL*** Bruce Coleman/John Cancalois, ***BR*** Corbis/Ecoscene; **53** Oxford Scientific Films/Rudie Kuiter; **54** ***T*** Hutchison Library/Lesley McIntyre, ***M*** Corbis/Papilio, ***B*** Oxford Scientific Films/Jeff Foott/OKAPIA; **56** ***M*** Oxford Scientific Films/Rafi Ben-Shahar, ***B*** Photodisc; **57** ***TM*** Wellcome Library, ***TR*** OSF/Peter Parks, ***M*** OSF/Peter Parks; **58** ***T*** Holt Studios/Phil Mitchell, ***B*** Woodfall Wild Images/David Woodfall; **59** Holt Studios/Nigel Cattlin; **60** ***T*** Getty Images/Stone, ***B*** Science Photo Library/Simon Fraser; **65** Photodisc; **67** ***T*** Mary Evans Picture Library, ***B*** Science & Society Picture Library/Science Museum; **72** Martin Stirrup; **73** Mary Evans Picture Library; **78** Peter Gould; **84** Peter Gould; **85** Peter Gould; **93** Photodisc; **94** Science Photo Library/Simon Fraser; **99** Photodisc; **100** Ace/PLI; **102** ***T*** Environmental Picture Agency, ***B*** Tony Gudgeon; **104** ***(both)*** Environmental Picture Agency; **106** Corbis/Adam Woolfitt; **107** Hutchison Library/Tony Souter; **108** ***T*** CEC/Mark Boulton, ***B*** CEC/Mark Boulton; **109** CEC/Mark Boulton; **110** Environmental Images; **111** Photodisc; **117** Photodisc; **118** ***T*** What PC, ***M*** Janine Weidel; **122** Repfoto; **124** ***TR*** Peter Gould, ***MR*** Art Directors & Trip, ***BR*** Garden Matters, ***BL*** Art Directors & Trip; **127** Trevor Clifford; **128** ***T*** Telegraph Colour Library, ***M*** Art Directors & Trip; **129** Art Directors & Trip; **130** ***T*** The Electricity Board, ***B*** Garden and Wildlife Matters Photo Library; **132** Ace/Dave Moore; **133** Morphy Richards; **134** ***T*** Peter Gould, ***B*** Art Directors & Trip; **136** ***T*** Science Photo Library, ***M*** Peter Gould; **138** Oxford Scientific Films; **139** Science Photo Library; **140** Science Photo Library; **142** ***T*** BNFL, ***B*** Science Photo Library; **144** ***T*** Oxford Scientific Films, ***B*** Eoscene; **145** Photodisc; **146** Rex Features; **153** Photodisc; **154** Ed.Pecheur d'images/Philip Plisson; **156** ***T*** CEC/Mark Boulton, ***M*** Science Photo Library/David Parker; **158** ***T*** Science Photo Library, ***BL*** Science Photo Library/Martin Bond, ***BR*** Science Photo Library/David Ducros; **159** ***T*** CEC/Mark Boulton, ***M*** Photodisc; **160** ***T*** Panos Pictures/Caroline Penn, ***M*** Science Photo Library/Pascal Goetgheluck, ***B*** Science Photo Library/Dave Roberts; **161** Mediscan; **164** Bruce Coleman/Kim Taylor; **165** Science Photo Library/Jesse; **168** Science Photo Library/David Nunuk; **169** Science Photo Library/James King-Holmes; **170** ***T*** Science Photo Library/NASA, ***M*** Photodisc; **172** ***T*** Science Photo Library, ***B*** Genesis Space Photography; **173** Science Photo Library/John Mead; **183** BNFL.

The publishers have made every effort to trace the copyright holders, but if they have inadvertently overlooked any, they will be pleased to make the necessary arrangements at the first opportunity.

Picture research by Thelma Gilbert.

# About this book

This is the first book in the Edexcel Modular Science series and is designed to guide you through units 1–6 of GCSE study. It has been written to help you find everything you need to prepare for your exams.

We have named each module in the book with the title of the module test that you will be set. This is so that you can be sure of what you need to study.

The introductory page to each module will help you get a feel for the module. It explains the topics covered and gets you thinking about the science that you already know.

To help you study we have included some useful features in the book. Here are a few:

**Double-page spreads**
Everything you need to know for each module is covered in double-page spreads. These pages will cover all the topics you need to understand for your exams. To help you find all of the important points we have included questions to test yourself as you learn.

**End of module questions**
At the end of each module you will find three pages of test questions. Answering lots of questions will help you check what you have learnt and prepare for exams.

**Did you know?**
These boxes contain extra information about the topic that you are studying. The information in these boxes will not be tested in module tests or exams.

**Key ideas**
The information in these boxes summarises the most important points on the page. In your exams you will be tested on your knowledge of the points in the Key ideas boxes. These boxes will also help you make notes and answer questions.

**Key words**
Important words are emboldened within spreads and listed at the back of the book to aid revision. These are ordered according to spread, so that you can check what you need to learn and easily find the page in which it is covered. Writing these out and filling in your own definitions will help you to prepare for your exams.

# Contents

# Module 1 – The human body: action and control

This module is divided into five parts.

1 The digestive system is where food is broken down and then absorbed into the bloodstream. Different parts of the digestive system have different conditions in order to achieve this.

2 The food molecules are transferred around the body in the bloodstream. Blood is made up of different components. Each component has a different job.

3 The nervous system controls the way the body operates. Messages are sent around the nervous system through nerves. The central nervous system (brain and spinal cord) is alerted to stimuli and brings about suitable responses.

4 The kidneys filter the blood to remove harmful chemicals and control the water level in the body.

5 The skin is the largest organ in the body. It has important jobs in defending the body from attack by harmful microbes, and in maintaining constant conditions.

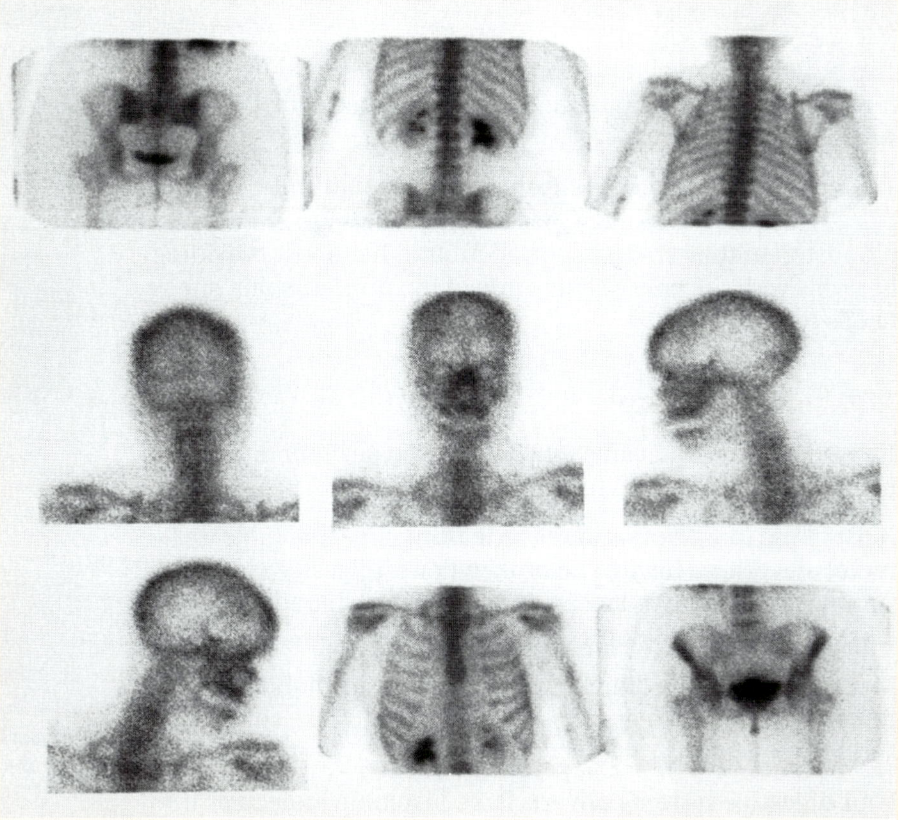

Try these questions first.

1 Why do we need a balanced diet?

2 A balanced diet contains seven different components. What are these?

3 Which component provides amino acids for building and repairing the body?

4 A runner has a large meal of pasta the night before taking part in a marathon. What does this meal provide as a good source of energy?

5 Why do children need a good source of calcium in their food? What is a good source of calcium?

6 Harry is told to eat a diet with low fat and high fibre. Which of the following forms of potato would be best for him?

chipped potatoes　potato crisps　jacket potatoes
mashed potatoes　boiled potatoes

# 1:1 Cutting food down to size

## Getting food into the body

The patient in the photograph opposite is getting the food molecules she needs through a drip connected directly into her bloodstream.

The drip contains soluble sugars and other molecules which can dissolve in her blood. Once the molecules are in her bloodstream they are transported around her body, providing her cells with the energy and some of the nutrients that she needs to recover.

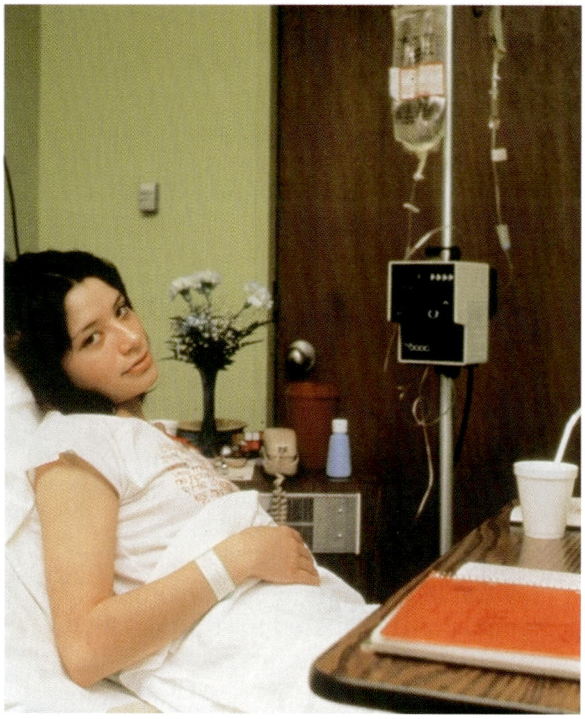

Sugars from the bag flow down a tube and pass directly into the girl's bloodstream.

## Making foods usable

If you are not ill and can eat normally, you get your energy and nutrients from the food in your diet.

Most of the foods you eat are made from large insoluble molecules. These have to be broken down (digested) into small soluble molecules which can dissolve in your blood and be carried around your body.

Many foods contain starch, which is a large insoluble molecule. Starch is digested into sugars, which are small soluble molecules.

Your **digestive system** converts large insoluble molecules into small soluble molecules.

**a Explain why hospital drips contain only soluble molecules.**

## Small molecules

Your diet contains three main types of food molecules – **carbohydrates, proteins** and **fats**. Each type of food molecule is insoluble and has to be digested into small soluble molecules. The table shows what is produced when each type of food is digested.

| Type of food | Product of digestion |
|---|---|
| carbohydrate | sugars |
| fats | fatty acids and glycerol |
| protein | amino acids |

**b A cheese sandwich contains protein (the cheese), carbohydrate (the bread) and fat (the butter). What is produced from each of these foods when they are digested?**

## Breaking down your food

Your digestive system produces **enzymes**. These **catalyse** (speed up) the breakdown of larger molecules into smaller ones. The diagram opposite shows how enzymes work to digest food.

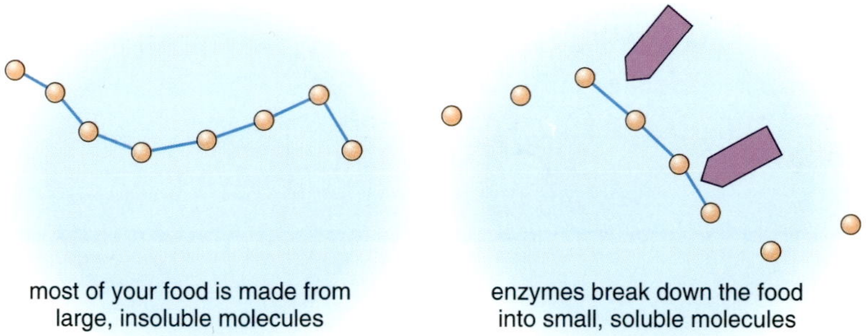

## The right tools for the job

Each type of food molecule needs a particular enzyme to break it down. For example, **protease** enzymes digest protein into amino acids. Protease enzymes digest only protein and will not break down the other types of food molecule.

The enzymes that catalyse the breakdown of carbohydrates into sugars are called **carbohydrases**.

**Lipase** enzymes catalyse the breakdown of fats into fatty acids and glycerol.

**c Give the names of three enzymes.**

The diagrams below show how enzymes work to digest your food into small soluble molecules which can be absorbed into your body and dissolve in your bloodstream.

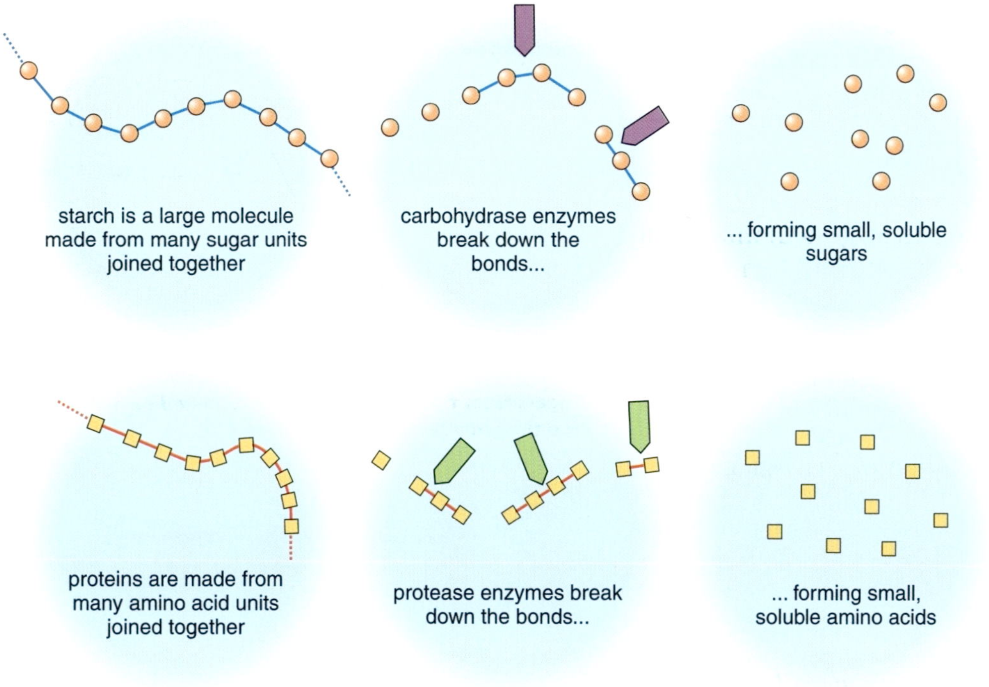


> **Did you know?**
> Enzymes have been used in baking bread and making wine for thousands of years.
>
> The word enzyme comes from Greek and means in yeast. It was first used by Wilhelm Kuhne in 1900.

## Questions

**1 Copy and complete this table.**

| Type of food | Type of enzyme | Digestion products |
|---|---|---|
| starch | | sugars |
| | lipase | |
| | | amino acids |

**2 Explain why proteins in the food you eat need to be broken down by enzymes.**

**3 Explain why health drinks such as Lucozade contain sugar.**

> ### Key ideas
> - *Starch, protein and fats are insoluble.*
> - *During digestion large insoluble molecules are broken down into small soluble molecules.*
> - *Only small soluble molecules can be absorbed into the bloodstream.*
> - *Digestion is speeded up by enzymes.*

# 1:2 Speeding things up

## Making big molecules smaller

Much of the food you eat is made from large insoluble molecules. But only small molecules can pass from your gut into your bloodstream. Large food molecules have to be broken down (digested) as they pass along the digestive system.

**a Why does your food need to be broken down?**

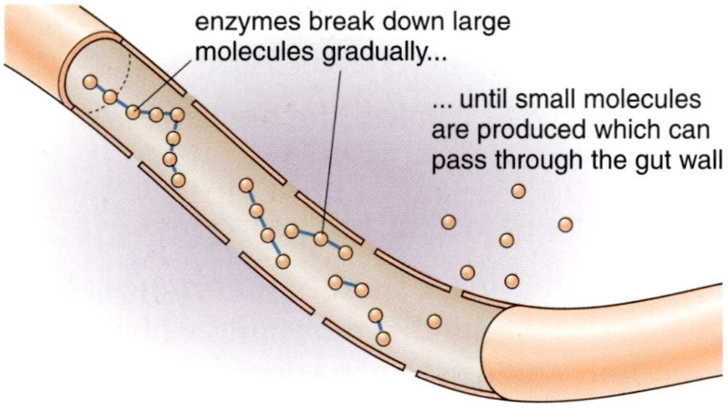


**Only small molecules can pass through the wall of the gut and dissolve in your bloodstream.**

## Stages of digestion

The diagram shows the main parts of the digestive system. Each part of the digestive system releases different enzymes onto the food you have eaten to break it down.

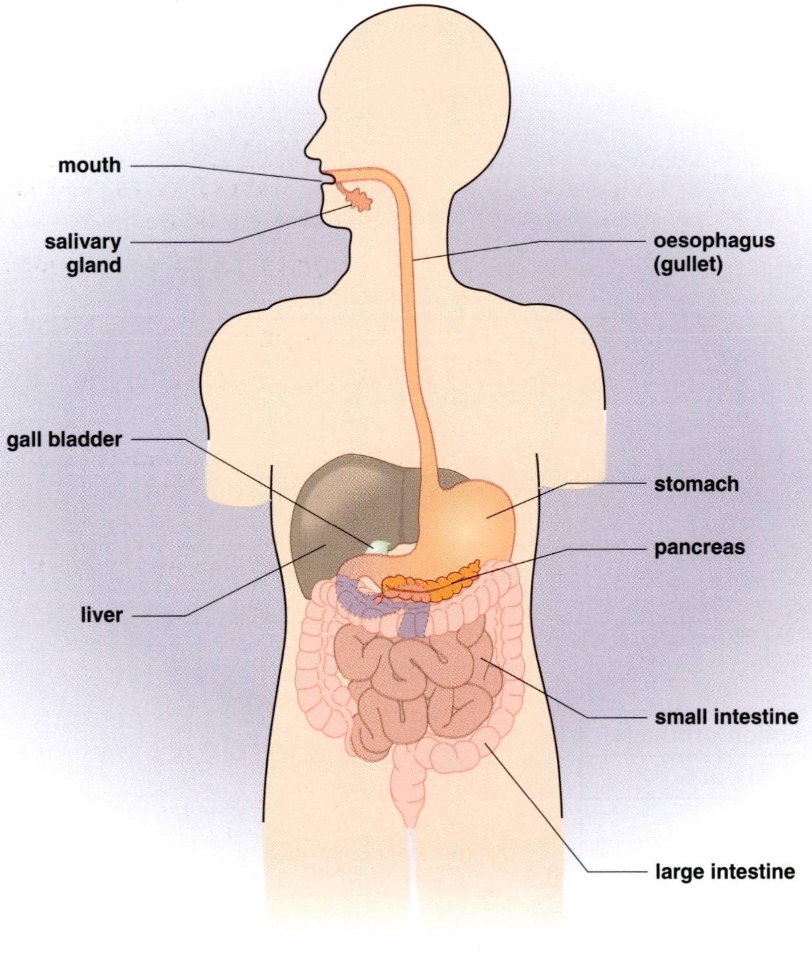


**Food is gradually broken down as it passes along the gut.**

**Did you know?**
If the digestive system of a human from mouth to anus was stretched out it would be several metres long.

The table opposite shows the jobs of the different parts.

**b How are conditions different in the stomach and small intestine?**

## Big drops to little droplets

Bile is made in the liver and is stored in the gall bladder. It is released from the gall bladder onto food in the small intestine.

The bile is alkaline. When it mixes with food, the bile neutralises the acid produced by the stomach and makes the conditions in the small intestine alkaline. All the enzymes released in the small intestine work best in alkaline conditions.

Bile also helps to digest fats. It does this by breaking down large drops of fat into thousands of tiny droplets. This process is called **emulsification**. The total surface area of the tiny fat droplets is much greater than the surface area of the large drop of fat. Emulsification by bile therefore increases the surface area of fats for lipase enzymes to act on and so speeds up the digestion of fat, as shown below.

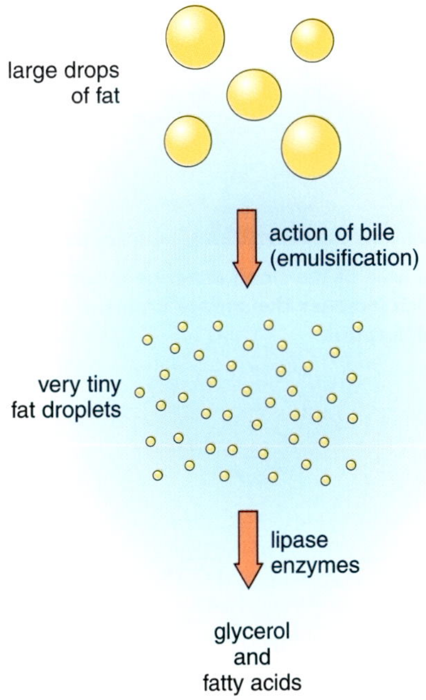


**Many fat droplets have a large surface area for lipase enzymes to work on.**

| Part of digestive system | Job |
|---|---|
| salivary glands | release carbohydrase enzyme into the mouth |
| mouth | food is broken down into smaller pieces by chewing (mechanical breakdown)<br>food is mixed with saliva<br>enzymes in the saliva start to break down the carbohydrate chemically |
| oesophagus (gullet) | the food slides down this tube to the stomach |
| stomach | food is mixed with gastric juices and acid<br>gastric juices contain protease which starts to break down the food<br>acid kills bacteria and provides the conditions needed by the protease |
| gall bladder | stores bile<br>bile is made in the liver and is involved in the digestion of fat |
| pancreas | produces enzymes such as protease, carbohydrase and lipase<br>these enzymes work best in alkaline conditions |
| small intestine | this is a long tube<br>digestion is completed here in alkaline conditions<br>small molecules from digestion are absorbed from here into the bloodstream |
| large intestine | only undigested food reaches here (see Spread 1:3) |

**c What sort of chemicals help enzymes to break down your food during digestion?**

## Key ideas

- *Each part of the digestive system has its own job in digesting food.*
- *Digestive organs also produce:*
  - *enzymes;*
  - *acid or alkaline substances so that enzymes work well.*
- *The liver produces bile, which neutralises the acid added to food in the stomach.*
- *Bile also emulsifies fats.*

## Questions

**1** Copy and complete the following sentences. Use words from the list.

| | | | |
|---|---|---|---|
| **fats** | **gall bladder** | **lipase** | **liver** |
| **protease** | **small intestine** | **starch** | |

Bile is produced in the __________ and stored in the __________. It is added to the food in the __________. When bile is added to food, it emulsifies __________ so that __________ enzymes can work better.

**2** What is the main job of bile?

# 1:3 Getting food into the body

## What happens to digested food?

The process of digestion is complete when food molecules have been broken down into small soluble molecules. The next stage is to get digested food into your body. This process is called **absorption**.

Molecules of digested food are small enough to be absorbed through the gut wall into the bloodstream. The wall of the small intestine acts like a sieve, allowing only small molecules to pass through it.

Digested food molecules are also soluble so that they will dissolve in the bloodstream. The blood transports the food molecules to all parts of the body.

The surface area of the small intestine is covered by villi, which increase the surface area and so speed up absorption.

## Getting food into the bloodstream

Your small intestine is very efficient at absorbing large amounts of food molecules. Its structure is specialised to carry out this job in the following ways.

- The inner surface contains tiny folds, and thousands of finger-like projections, called **villi** (singular: **villus**). These create a very **large surface area** for food molecules to pass through. The villi are shown in the photograph.
- Each villus contains many tiny blood vessels (capillaries) to transport absorbed food molecules.
- Each villus is very thin, so food molecules quickly reach the bloodstream.

The diagram below shows how these features allow absorption to take place very efficiently.

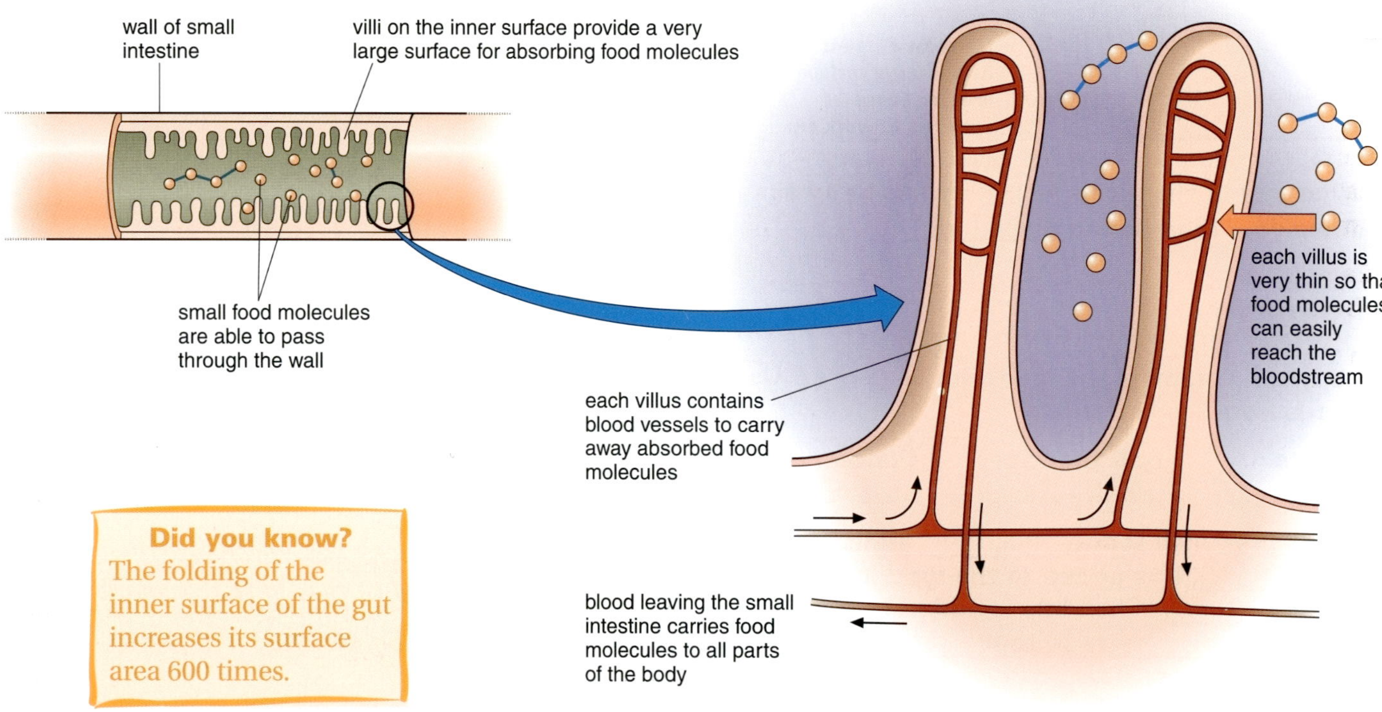


**Did you know?**
The folding of the inner surface of the gut increases its surface area 600 times.

**a List three features that make villi efficient at absorbing the products of digestion.**

## Getting food to all parts of the body

Once food molecules reach the bloodstream they dissolve in plasma – the liquid part of the blood. Your gut has a large blood supply in order to transport all the absorbed food molecules away from the gut. This is why it is important to have such a large number of blood capillaries in the wall of the small intestine.

**b List three food substances that are likely to be found in the plasma of blood leaving the small intestine.**

## Getting rid of waste

When food reaches the large intestine all the soluble food molecules will have been absorbed, leaving behind substances that cannot be digested, such as fibre (cellulose). Even though you cannot digest the fibre, it still has benefits. Fibre helps to prevent constipation and diseases of the intestine, including cancer.

In the large intestine much of the water is absorbed from the undigested material, leaving behind semi-solid waste called faeces. This waste is then removed from the body through the anus.

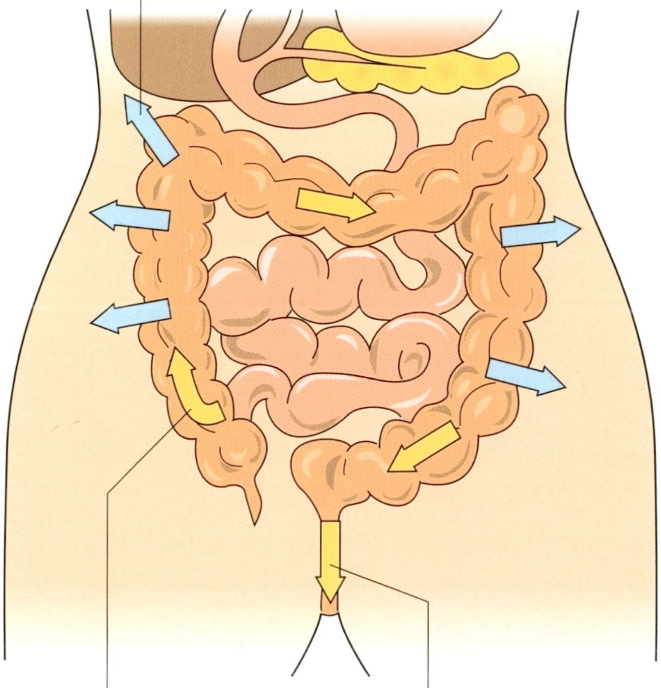


**The large intestine absorbs water and forms waste (faeces).**

## Questions

**1** Copy and complete the following sentences. Use words from the list.

**anus** **blood vessels** **faeces** **specialised** **surface area** **villi** **water**

The structure of the small intestine is ____________ to carry out its job. The inner surface contains thousands of tiny projections called __________. These give the inside of the small intestine a large ________________.

Each villus contains tiny __________ to transport the soluble products of digestion.

In the large intestine __________ is removed from the undigested food and is re-used. The undigested food is lost from the body as ____________ through the __________.

**2** Many breakfast cereals are described as 'high fibre'. What are the benefits of including these in the diet?

**Did you know?**

Dieticians recommend a low-fat and high-fibre diet for most people. Some breakfast cereals are high in fibre.

## Key ideas

- *The soluble products of digestion are absorbed through the wall of the small intestine into the bloodstream.*
- *The small intestine has special features which make it very efficient at absorbing the products of digestion.*
- *The large intestine absorbs water from the indigestible food which forms faeces.*

# 1:4 Supplying the body's needs

## Life blood

Your blood provides all the cells of your body with the materials they need. It also removes waste substances. Cells would soon stop working without a good blood supply.

If people have lost a lot of blood in an accident they can be given blood that has been donated by someone else. Blood transfusions are also given to treat some diseases. The blood being given to the patient in the photograph will provide him with the new cells that he needs.

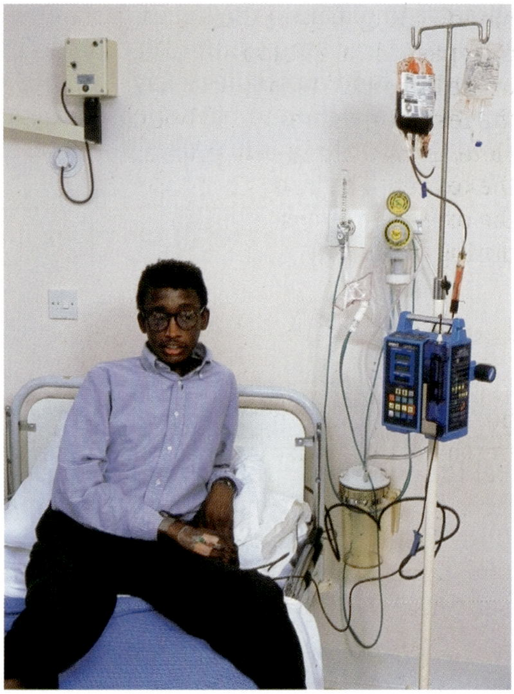

Transfusions of donated blood save lives.

## What is in your blood?

Your blood contains a large number of cells floating in a liquid called **plasma**. The diagram and photograph below show what the cells in your blood look like and what they do.

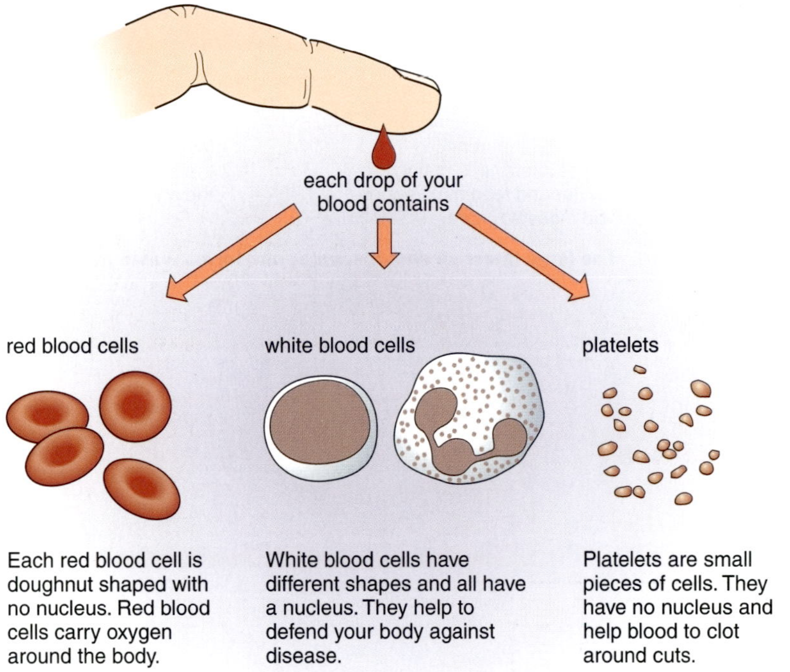


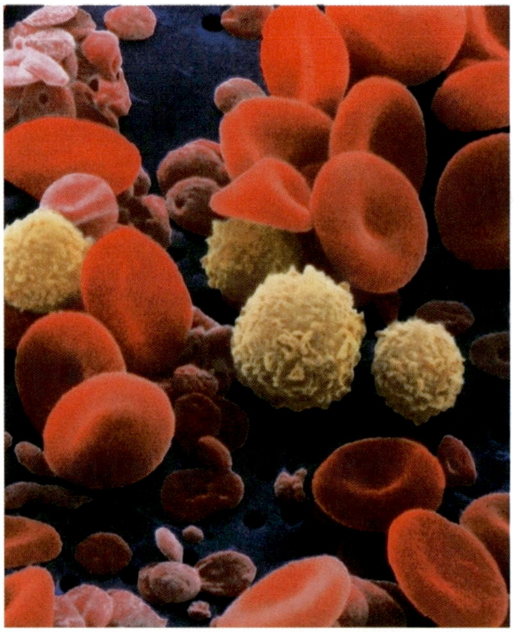

A photograph of blood cells, taken using a microscope.

**a Which part of the blood fights infection?**

**b Which part of the blood helps blood to clot?**

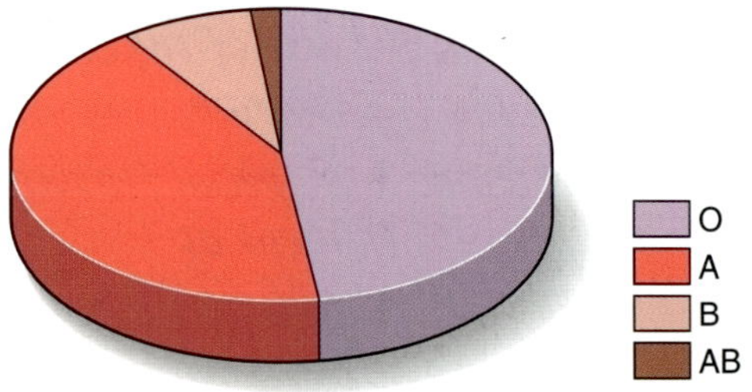


**c Which is the commonest blood group?**

### Did you know?

We all have one of four blood groups: A, B, AB or O.

The pie chart shows the percentage of people with each blood group.

## Carrying oxygen

Blood contains an enormous number of red blood cells. Red blood cells transport oxygen from your lungs to all the cells of your body.

The diagram shows how the red blood cells pick up oxygen as they pass through blood vessels in the lungs and transport it to the respiring tissues of the body. Here, the red blood cells give up the oxygen. The body's cells use the oxygen to release energy during respiration.

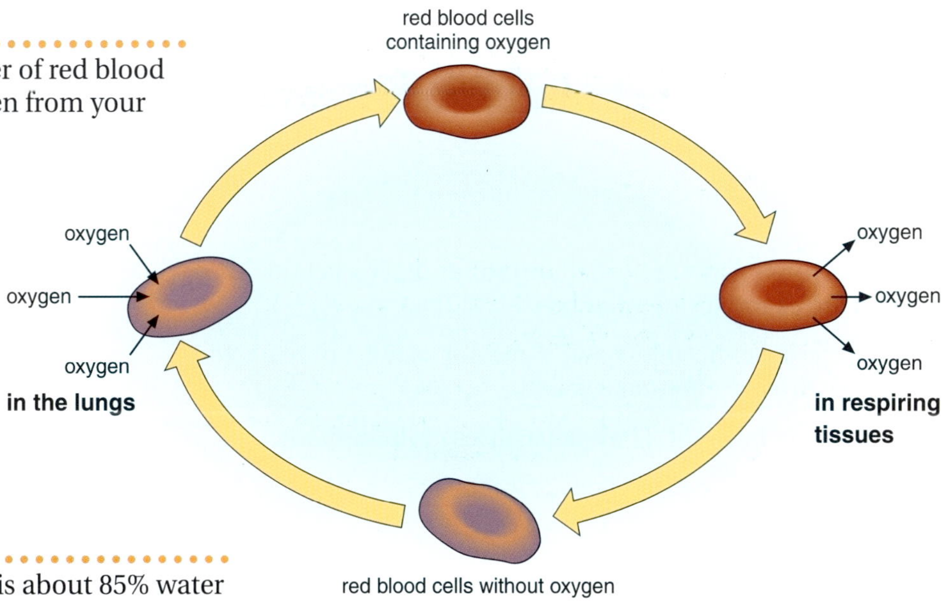


## Transporting useful materials

Plasma is the liquid part of blood. It is about 85% water but has many substances dissolved in it. Soluble substances are transported in plasma from one part of your body to another.

- **Carbon dioxide** is produced by all body cells as they respire. The carbon dioxide produced by body cells is carried in plasma to your lungs, where it is breathed out.
- **The soluble products of digestion** (e.g. glucose and amino acids) are absorbed from the small intestine and transported in plasma to other parts of the body.
- **Urea** is a waste substance produced in your liver. It is transported in plasma, from the liver to your kidneys, where it is removed from your body in the urine.
- **Hormones** produced by various glands in the body are transported in the blood to their target organs, where they have their effects.

**Did you know?**

There are about 5.5 million red blood cells in each cubic centimetre of blood. The bone marrow has to constantly produce red blood cells because they have an average life span of only 120 days.

## Questions

**1** **List four substances transported in blood plasma.**

**2** **Copy and complete the following table.**

| Type of blood cell | Appearance/function |
|---|---|
| | transports oxygen around the body |
| | helps the blood to clot |
| | contains a nucleus |
| | are biconcave in shape |
| | defends the body against disease |

**3** **Copy and complete the following table by writing true or false in the spaces.**

| Plasma transports … | |
|---|---|
| oxygen from the lungs to muscles | |
| starch from the small intestine to muscles | |
| carbon dioxide from muscles to the lungs | |
| carbon dioxide from the lungs to muscles | |
| urea produced by the kidneys to the liver | |

### Key ideas

- *Blood transports useful and waste substances around the body.*
- *Blood contains red blood cells, white blood cells and platelets. Each type of cell carries out a different function.*
- *Soluble substances are transported in plasma.*

# 1:5 Microscopic enemies

## Harmful microbes

Very small organisms are called **microbes**. **Bacteria** and **viruses** are examples of microbes.

Most bacteria are harmless and some are useful. Some bacteria can cause disease and make you ill.

All viruses are harmful. They cause diseases in humans, animals and plants.

**Did you know?**
Bacteria are used to clean up waste water at sewage works.

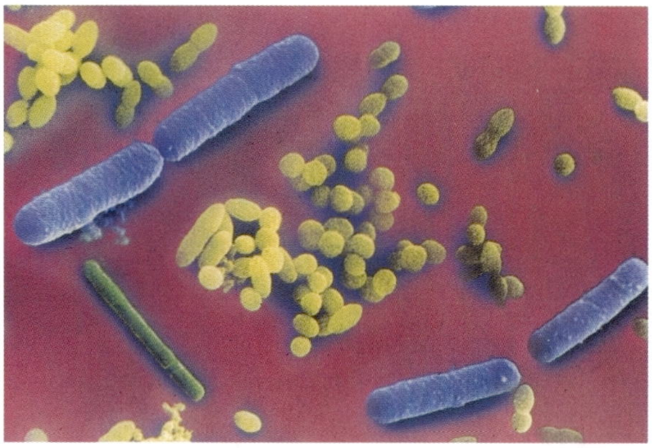

Bacteria are tiny single-celled organisms.

## Tiny invaders

The bacteria that can enter your body and cause disease are very, very small. They are smaller than your body cells and can only be seen with a powerful microscope.

The diagram opposite compares a bacterial cell with a typical cell in your body.

Both cells contain **genes**. Genes contain information that controls what happens in a cell. The genes are found in **chromosomes**. In human cells chromosomes are found inside the nucleus. The chromosomes of bacteria are not in a nucleus.

**a** Which two features from the list below are in a bacterial cell and a human cell?

**b** Which feature is in a bacterial cell but not in a human cell?

cell membrane    cell wall
cytoplasm    nucleus

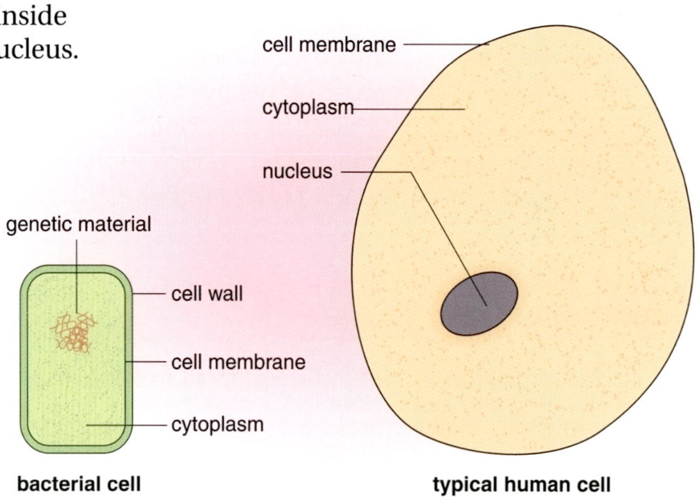


A bacterial cell and a human cell magnified about 2000 times.

## Rapid reproduction

Bacteria can reproduce very quickly. Bacteria reproduce by dividing into two. If the conditions are right, bacteria can divide every 20 minutes. This means that a small number of bacteria can increase to thousands in a very short time.

**c** If 100 bacteria get into your body and reproduce every 20 minutes, how many will there be in 2 hours? Copy and complete the table to work this out.

| Time (min) | Number of bacteria |
|---|---|
| 0 | 100 |
| 20 | 200 |
| 40 | 400 |
| 60 | |
| 80 | |
| 100 | |
| 120 | |

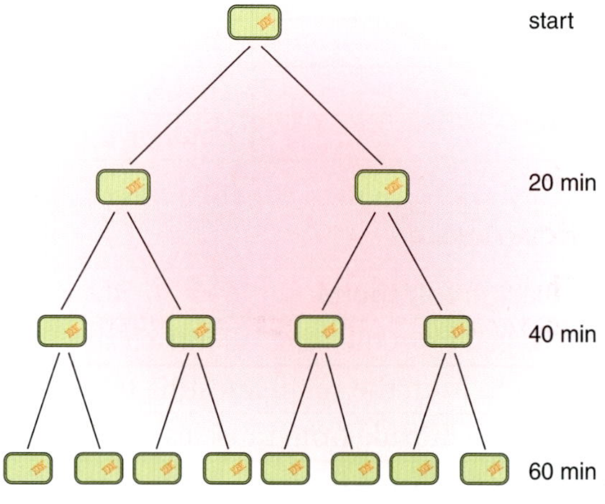


The number of bacteria doubles after each division.

## The smallest invaders

Viruses are even smaller than bacteria. The diagram opposite shows one type of virus. Viruses have a coat made of protein. This coat surrounds a few genes.

Viruses cannot reproduce by themselves. They can only reproduce inside living cells such as human cells. Viruses invade living cells and, once inside, use the host cell to make thousands of new viruses. The new viruses burst out of the host cell, which damages the cell. The new viruses then invade other cells.

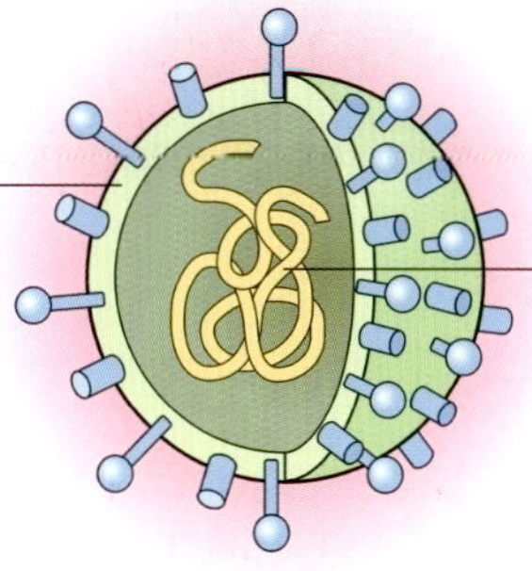


**A virus magnified many times.**

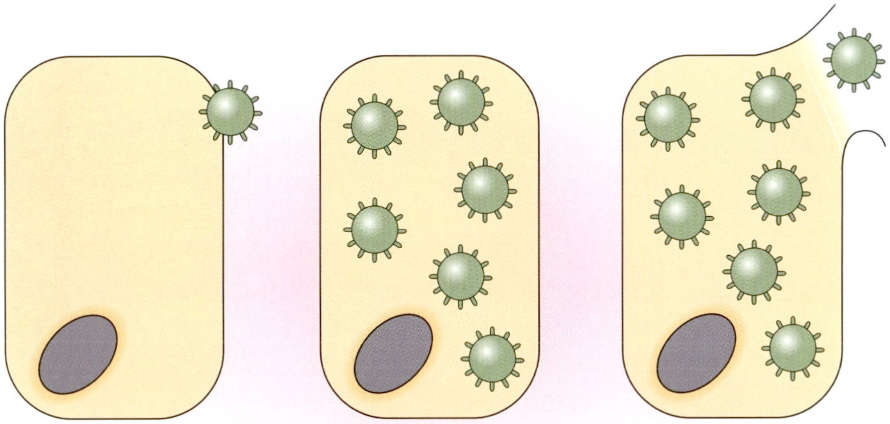


**How a virus reproduces inside a host cell.**

### Did you know?

Antibiotics kill bacteria but have no effect on viruses.

The first antibiotic, penicillin, was discovered by accident in 1928 by Sir Alexander Fleming. However, it took 15 years for scientists to find a way to produce it in a form that could be used to treat patients. Now the lives of millions of people have been saved by penicillin and other antibiotics.

A patient with bacterial meningitis is treated with a cocktail of nine different antibiotics.

## Questions

1 Copy and complete the following sentences. Use words from the list.

**chromosomes genes nucleus protein reproduce respire wall**

The cells of bacteria contain cytoplasm, a cell membrane and a cell _____. Their chromosomes are not in a ________.

Viruses can only _________ inside living cells. Viruses have a ______ coat surrounding a few ______.

2 How do bacteria reproduce?

3 List three differences between bacteria and viruses.

4 How does a virus damage cells?

5 There are different types of meningitis. Why can't doctors use antibiotics to treat viral meningitis?

## Key ideas

- *Viruses and bacteria are examples of microbes.*
- *A bacterial cell has cytoplasm and a membrane surrounded by a cell wall.*
- *The genes in a bacterial cell are not in a nucleus.*
- *Viruses have a protein coat that surrounds a few genes.*
- *Bacteria and viruses can reproduce very quickly inside your body.*
- *Viruses must invade living cells in order to reproduce.*
- *Viruses damage the cells in which they reproduce.*

# 1:6 Spreading diseases

## Catching a disease

Many diseases are caused when microbes such as viruses and bacteria get into the body.

You are more likely to develop a disease if large numbers of viruses or bacteria enter your body. This can happen by being in unhygienic conditions or by being in contact with someone who already has a disease.

Once viruses or bacteria are in your body they reproduce rapidly. There are many places inside you where bacteria and viruses can live and reproduce.

Some bacteria release poisons called **toxins** which make you feel ill.

**a Look at the cartoon. Why should you not buy a hot dog in this cafe?**

Would you buy a hot dog here?

## Spreading harmful microbes

Microbes that cause disease can be spread around in many ways. Here are just a few examples.

When someone coughs or sneezes, they release tiny droplets of moisture into the air. Microbes can stick onto the droplets and can get into your body when you breathe in. Common diseases such as colds and flu are spread in this way.

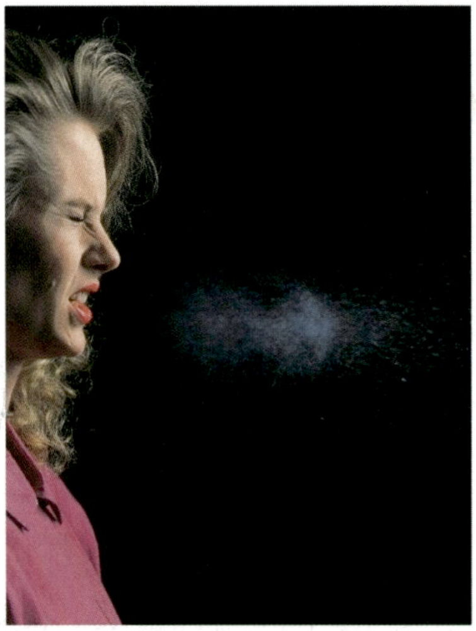

Coughs and sneezes spread diseases.

It is important to prepare and keep food in hygienic conditions.

Food poisoning is one of the most common illnesses. You become ill by eating food that contains certain harmful bacteria. This is why it is important to store and handle food in hygienic conditions.

Insects and other animals such as rats can spread harmful microbes. Houseflies spread bacteria from one place to another, as you can see in the poster opposite.

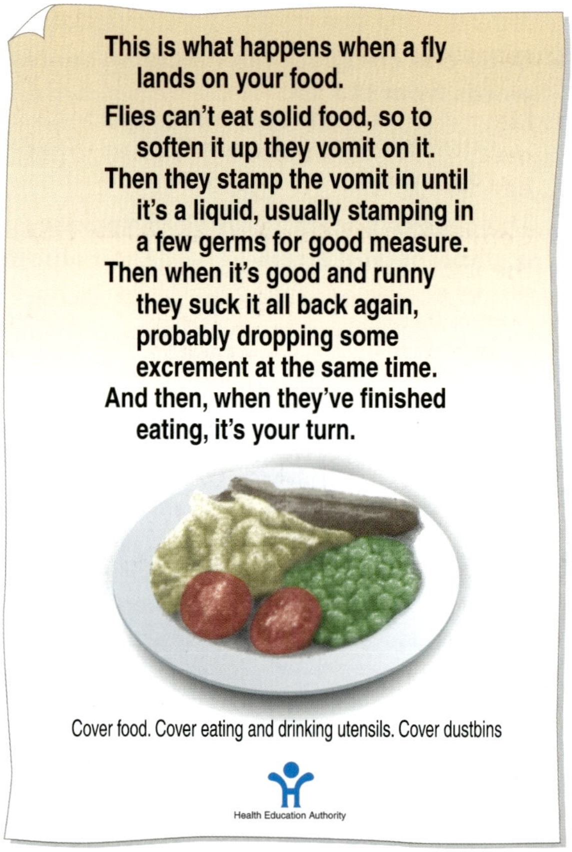

**b** Alice travels to school every day on a crowded bus. Chris walks to school. Suggest why Alice is more likely than Chris to catch a cold.

## Major disasters

Major disasters such as the floods that happened in Mozambique in February 2000 can kill hundreds of people.

People still faced danger even when the floods began to disappear. This is because diseases such as cholera can spread quickly following disasters. Cholera is caused by bacteria which are spread in water that has been contaminated with human sewage. In developing countries, a toilet may be a deep hole dug in the ground. When the ground floods, bacteria in the sewage are carried in the flood water.

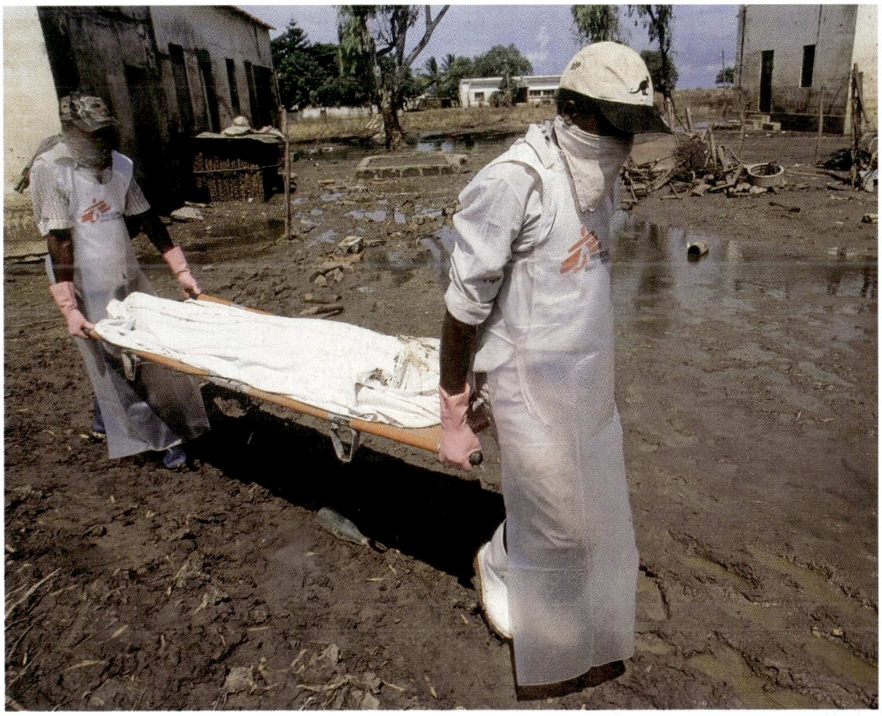
Rescue workers clearing up after floods must wear protective clothing to avoid infection.

## Stopping the spread of disease

In the past, living conditions made it easy for microbes to breed quickly. In early towns and cities there were no drains or rubbish collectors. Decay in the streets provided the perfect conditions for microbes to reproduce and spread.

Modern towns and cities are designed to reduce the spread of disease.

### Did you know?

In the nineteenth century the river Thames was particularly unhealthy. The year 1858 was called 'The Year of the Great Stink'. A huge programme to build sewers followed.

In 1854 John Snow showed that polluted water caused cholera. The last major outbreak of cholera in the UK was in 1937, when 43 people died in Croydon, Surrey.

## Questions

1 List three ways in which people handling food such as meat in shops can prevent the food from becoming infected with bacteria.

2 Complete the following sentences. Use words from the list.

| | | |
|---|---|---|
| animals | bacteria | contaminated |
| disasters | microbes | moisture |

Cholera is a disease caused by ________. People can develop this disease by drinking ________ water.

Flu is a common disease caused by ________ which is spread on ________ in the air.

3 Explain why it is important to cover food to keep flies off.

## Key ideas

- *You are likely to develop a disease if large numbers of microbes enter your body.*
- *Microbes reproduce when they get inside the body.*
- *Some microbes release toxins.*
- *Microbes can easily spread.*

# 1:7 Body defences

## Outer protection

Your body is always coming into contact with harmful microbes without you becoming ill. This is because your body is very good at keeping microbes out.

Your skin is a barrier that stops microbes getting inside your body.

**a Where can bacteria get into the body?**

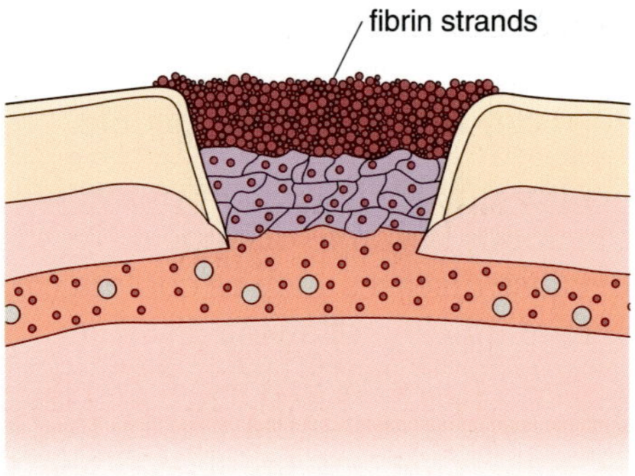


**Fibrin strands tangle together and plug the cut skin.**

## Sealing up a cut

If the skin is cut, microbes may get in and cause infection. To stop this happening, blood forms a **clot** that seals the cut.

The platelets in the blood release a chemical. This chemical changes and forms strands of fibrin. The strands tangle together and plug the cut.

The plug is a dark colour because red blood cells are trapped in it. The plug dries out to form a scab. The scab protects the skin whilst new skin grows underneath.

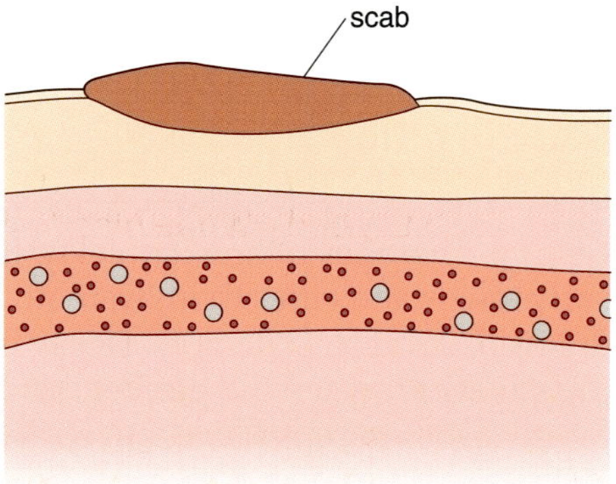


**A scab forms, protecting the new skin growing over the wound.**

## Trapping microbes

Harmful microbes can enter your body in the air you breathe.

Your breathing passages are lined with cells which produce a sticky liquid called mucus. Bacteria that are breathed in get stuck in the mucus.

Other cells lining your breathing passages have tiny hair-like projections called cilia. The movement of the cilia sweeps any trapped bacteria back up to your throat where they can be swallowed.

**b Cigarette smoke stops the cilia of the cells that line your breathing passages from working. Explain why smokers are more likely than non-smokers to develop throat infections.**

## Recognising harmful microbes

White blood cells can recognise harmful microbes from their surface features. The chemical structures on the surface of invading microbes are called **antigens**. The white blood cells then produce antibodies to destroy the microbes. White blood cells are part of the immune defence system.

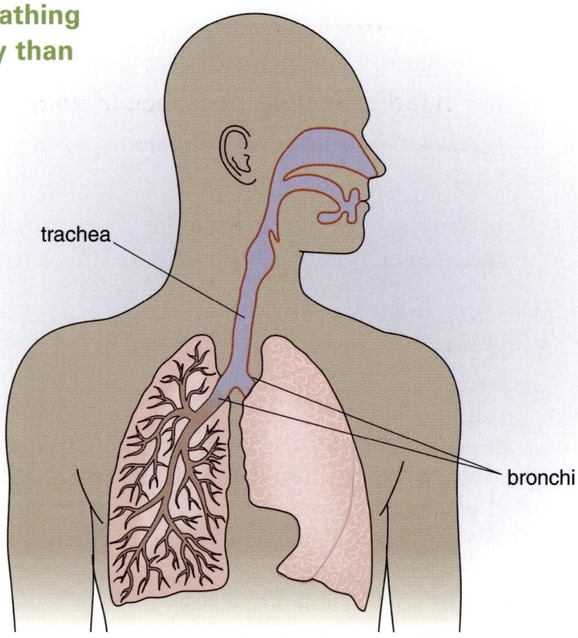


**Key**
— the lining of your breathing passages produces mucus to trap bacteria

> **Did you know?**
> The human immunodeficiency virus (HIV) attacks the immune system. Although a person can carry HIV for a long time without obvious effect, the virus can become active. HIV gradually destroys the immune system so that the person is less able to defend themselves against other infections. This leads to AIDS (acquired immune deficiency syndrome).

## Cells to fight diseases

If microbes do get into your body you are protected by your white blood cells. These cells are a very important part of your body's defence system against microbes.

There are two main types of white blood cell. They look similar but work in different ways.

- **Phagocytes** surround bacteria and digest them. The diagram at the top of the page shows how.
- **Lymphocytes** make proteins called **antibodies**, which help to kill invading bacteria. Antibodies are specialised to attack specific bacteria. The process is summarised in the diagrams opposite. When new bacteria enter the body, white blood cells make new antibodies. After the bacteria have been destroyed, some 'memory' lymphocytes remain to guard against re-infection.

White blood cells are squashy so they can squeeze along the tiniest blood vessel to where they are needed.

Some bacteria produce **toxins**. These are chemicals that poison the body. White blood cells also produce **antitoxins**. These neutralise the effects of the toxins.

**c Old people and people at risk can have a flu vaccination each year. The injection contains flu antibodies. Suggest why this can prevent the person from getting flu. Why doesn't this protect the person against a new strain of flu?**

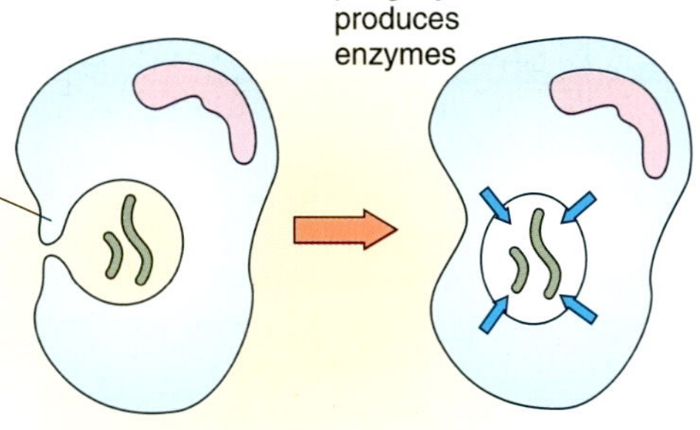


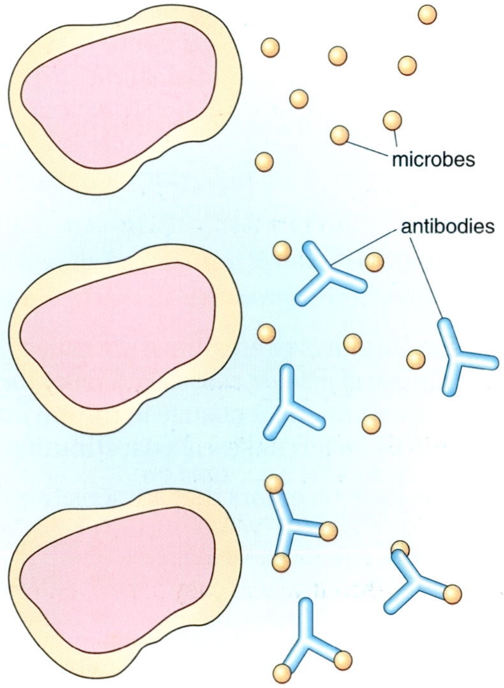


## Questions

1 Describe three ways in which white blood cells fight microbes in your body.

2 Copy and complete the following sentences. Use words from the list.

| | |
|---|---|
| antibodies | antitoxins |
| digest | surround |

Some white blood cells take microbes into their cells and ______ them. Other white blood cells destroy microbes by releasing _________. They can also release ______, which prevent the toxins produced by microbes from working.

3 Explain the difference between antigens, antibodies and antitoxins.

### Key ideas

- *Microbes are kept out of your body by your skin, blood clots and the mucus produced by cells lining the breathing passages.*
- *Blood clots are formed when a network of fibrin strands is formed.*
- *White blood cells are able to squeeze through the tiniest blood vessels.*
- *White blood cells destroy microbes by digesting them and by producing antibodies.*
- *White blood cells also release antitoxins.*

# 1:8 Detecting changes

The world around us is always changing. Animals need to detect these changes in order to survive.

## Receptors

The animal opposite is called the loris. The loris lives in tropical forests. It sleeps during the day, but at night it hunts for the small insects that it feeds on.

**a Why do you think the loris is successful hunting at night?**

The loris detects its prey by sight. Sharks also need to detect their prey, but they have very poor eyesight. However, sharks are able to detect a wounded animal from a great distance.

**b How do you think sharks detect wounded animals?**

Impala are prey. To avoid being eaten, they need to detect leopards and other predators.

**c How do you think impala can detect leopards that are crawling through grass towards them at night?**

The surroundings of an animal are called its **environment**. Animals use specialist cells called **receptors** to detect changes in the environment. A change in the environment that can be detected by an animal is called a **stimulus** (plural: **stimuli**).

Humans have receptors that are sensitive to:

- light
- sound
- touch and pressure
- chemicals
- changes in temperature
- changes in position

These receptors are found in different parts of the body. A part of the body containing many receptors is called a sense organ. Some of the body's sense organs are shown below.

**d Which of the sense organ(s) in the diagram contain(s) mainly receptors sensitive to:**

**i sound;**
**ii chemicals;**
**iii light**
**iv touch and pressure?**

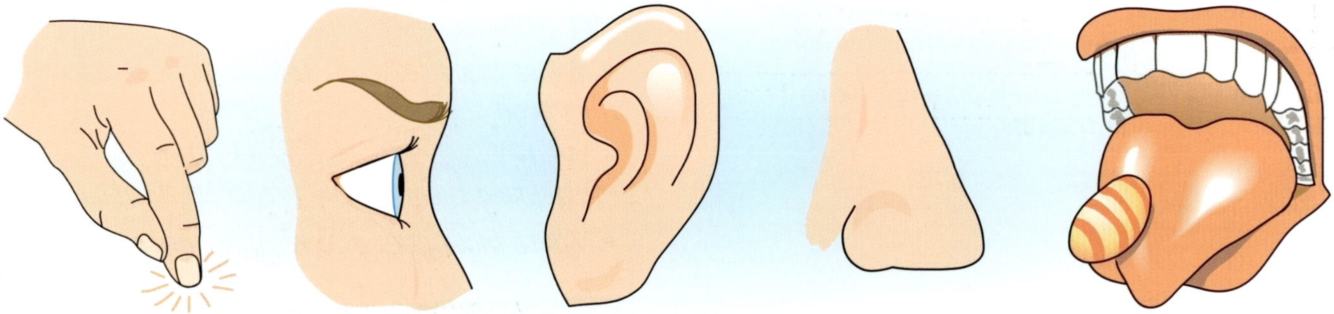

If pressure receptors are over-stimulated then we feel pain. That is why a tap does not hurt but a smack does.

As well as having receptors sensitive to sound, the ears have receptors that are sensitive to changes in the position of the head. These receptors help you to keep your balance. (It is these receptors that are affected when you spin round quickly and become dizzy.)

## The nervous system

Receptors only detect stimuli. The information from the receptors must be passed to the brain. Information is carried round the body by specialised cells called nerve cells. Nerve cells are gathered together to form **nerves**.

The brain and the spinal cord (which make up the **central nervous system**) consist of millions of nerve cells. Nerve cells from parts of the body below the head enter and leave the spinal cord, rather than the brain itself. The spinal cord carries the information to and from the brain.

The central nervous system and the nerves work together to form the **nervous system**. The brain and spinal cord are **co-ordinators** – they collect information from receptors and pass it on to the appropriate organs for action.

Information is passed around the nervous system by **electrical impulses**, which travel along the nerve cells.

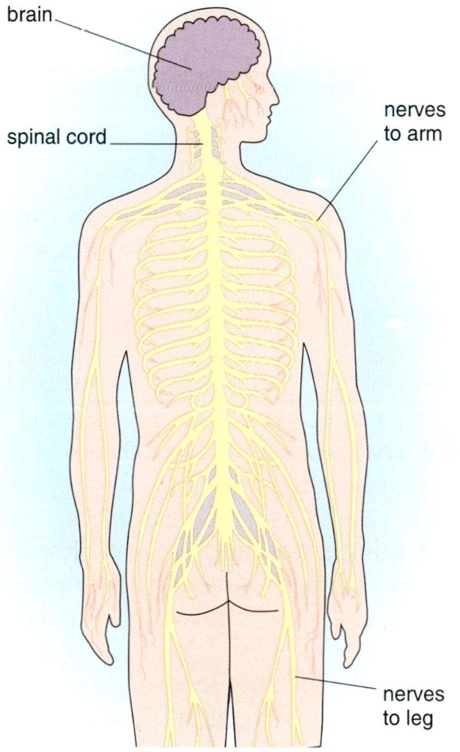

**Information is carried around the body by specialised nerve cells.**

## Questions

**1** Copy and complete the following sentences. Use the words in the list.

**brain** **nerve cells** **receptors**
**spinal cord** **stimulus**

A change in the environment is called a ____________. Animals have ________________ to detect these changes. Information is carried around the body by ________________ . The nervous system consists of nerves, the __________ and the ________________.

**2** List the main sense organs in the body.

**3** Copy and complete the following table.

| Receptor | Stimulus or stimuli |
|---|---|
| eye | ____________ |
| ear | ____________ and pressure |
| tongue | ____________ |
| skin | ____________, pressure and temperature |

### Key ideas

- *Cells called receptors detect stimuli.*
- *There are receptors in the eyes, ears, tongue, nose and skin.*
- *The nervous system consists of nerve cells.*
- *The brain and spinal cord co-ordinate responses.*

# 1:9 Reflex actions

The brain does a very important job in the nervous system. Information is passed along the nerves to the brain whenever a stimulus is detected. The brain makes sense of the information and co-ordinates a response.

## Automatic reactions

It takes time for the messages to reach the brain and a response to be returned to the appropriate organ. Sometimes the body needs to react more quickly than this. For example, if you put your hand on a hotplate it is important that you move it away very quickly. Severe pain and heat detected by receptors can cause a **reflex action** to take place. This is an automatic reaction – we do not need to think about it. The knee-jerk reflex is another example of a reflex action.

### The knee-jerk reflex

The knee-jerk reflex is used to test a patient's nervous system. The doctor taps the tendon just below the knee with a rubber hammer. If the knee jerks upwards automatically, the nervous system in the leg is working correctly. The diagram shows how the nervous system makes this happen. The patient's leg responds very quickly. We are born with reflex actions. The brain does not need to be involved in reflex actions.

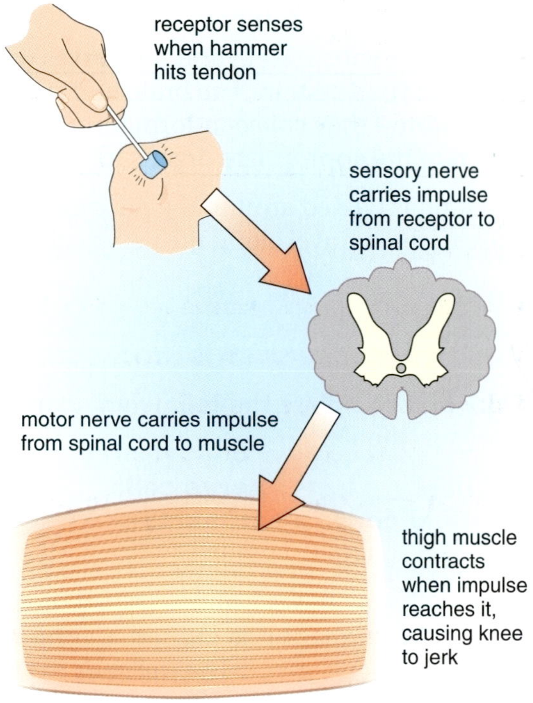


**The knee-jerk reflex.**

The brain only needs to become involved with actions that need decisions. Some reflex actions involve parts of the head rather than the body. In these cases, impulses are sent from sensory neurones to motor neurones that connect in the brain rather than in the spinal cord.

The brain and spinal cord are known as **co-ordinators** when they take part in reactions. This is because they cause impulses to pass to the appropriate neurones.

**a State whether each of the following actions is a reflex action or something we have learned to do:**

- **i** **blinking when dust lands on the eye;**
- **ii** **catching a tennis ball;**
- **iii** **jumping in response to standing on a pin;**
- **iv** **sneezing;**
- **v** **putting your books away as soon the bell rings.**

## Glands in reflexes

Not all our reflexes depend on muscle movement. Other reflexes help the body to work properly. If you are queuing for school lunch and you smell food that you like, your mouth will begin to water. In this case it is not a muscle that brings about the reaction but glands – the salivary glands in the mouth. This is the mouth-watering reflex.

## Questions

1 Copy and complete the following sentences. Use words from the list.

**automatic** **nerves** **receptor**

A reflex action is an ________ reaction to a stimulus. The stimulus is detected by a ________.

Electrical impulses travel towards the brain or spinal cord via __________ .

2 The following table gives some features of a nerve cell and why this makes the cell able to send messages around the body.

Copy and complete the table.

| Feature | Why this helps to send messages |
|---|---|
| very long | |
| has many projections which receive impulses | |
| has a fatty sheath | |
| the end of the cell branches many times | |

3 If you step on a pin, you automatically lift your foot away from the pin.

a What type of receptor is sensitive to the pin?

b What sort of reaction is this?

4 A doctor tests the knee-jerk reflex of a patient.

a How does the doctor do this?

b What does the doctor expect to see?

c Describe what happens to the body when the doctor tests the knee-jerk reflex.

5 The list below contains four reflex actions:

a knee-jerk reflex;

b mouth-watering reflex;

c blinking reflex;

d reflexes in the feet.

Which are controlled by the brain, and which by the spinal cord?

**Did you know?**

Hiccups are caused by a reflex action. If you eat too much, the pressure of your stomach on the diaphragm causes a reflex action.

## Key ideas

- *Nerve cells are very long cells so they are good at transporting impulses around the body.*
- *Nerve cells are gathered together to form nerves.*
- *In a reflex action, impulses from a receptor pass along a sensory neurone to the central nervous system.*
- *Impulses are then sent along the motor neurone to the effector, which brings about the response.*
- *In reflex actions the brain is not involved in a conscious way in deciding the response.*

# 1:10 The eye

As we get older the lenses in our eyes sometimes become cloudy and we cannot see clearly. The cloudy region of a lens is called a cataract. Every year, in the UK alone, thousands of people partially lose their sight because of cataracts.

Doctors can replace the natural lens with a plastic lens that works just as well, shown in the photograph. This is one of the most successful operations performed today.

**a What is the job of a lens?**

The diagrams below show the parts of the human eye.

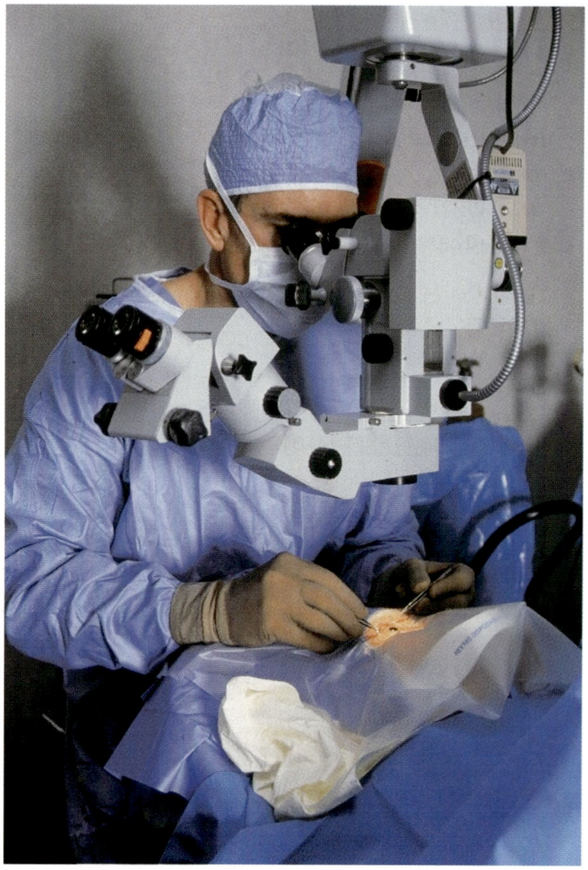

## Seeing in the dark and in the light

Light enters the eye through the pupil. Very bright light can damage the retina, which is why you must never look directly at the Sun. The eye can reduce the amount of light entering in very bright conditions.

The size of the pupil changes according to the light conditions. In bright light, circular muscles in the iris contract and the pupil gets smaller. This means that less light is let into the eye. In dim light, radial muscles in the iris contract so that the pupil gets larger and more light enters the eye.

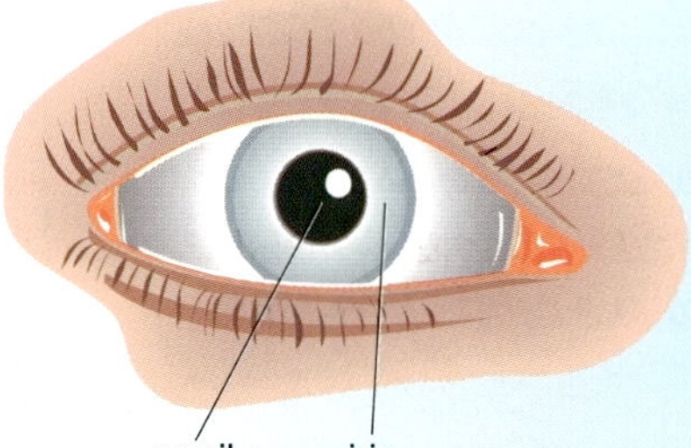


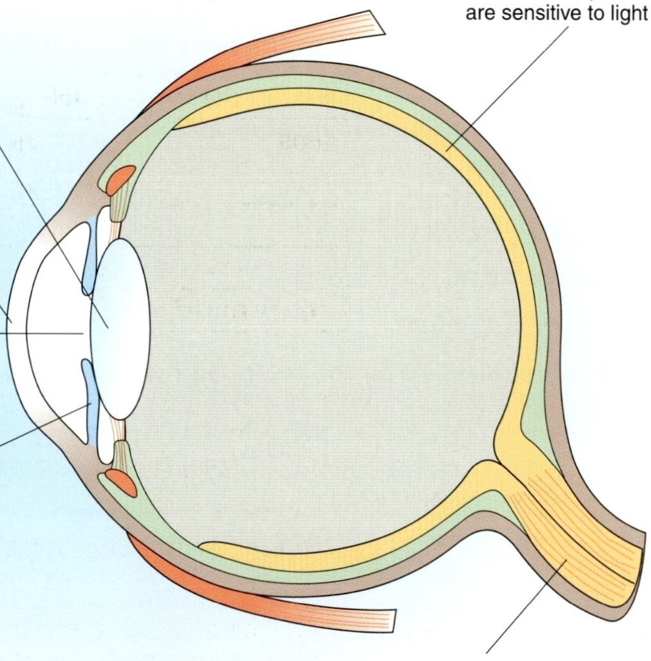

**b** Look at the pictures on the right. One shows an eye in bright light and the other an eye in dim light. Which is which? Explain your answer.

**c** Is this change a reflex action or do you have to plan to do it?

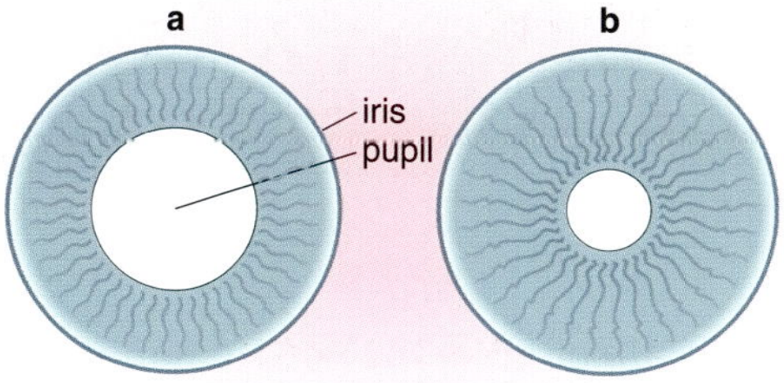


## Sending information to and from the brain

The back of the eye contains receptor cells that are sensitive to light. This layer of cells is called the **retina**. The light receptor cells are connected to sensory neurones. These sensory neurones carry electrical impulses from the light receptors to the brain. The sensory neurones are grouped in the **optic nerve**. The brain makes sense of the impulses.

Tha amount of light entering the eye is called the iris reflex. it can be summarised as:

**receptor on retina** → **impulse sent through optic nerve** → **brain** → **impulse sent** → **iris muscles respond**

## Cones and rods

When we look at an object, the light rays from the object are focused on to the retina. If the focus is not on to the retina, the image is blurred. Lenses can be used to correct this.

On the retina there are two types of detector:

- rods – these detect black and white;
- cones – these detect colour.

**Did you know?**
A detached retina can be repaired by a routine laser treatment.

## Questions

1 List in order the structures that light passes through on its way from the air to the retina.

2 Copy and complete the table.

| Part of the eye | Job |
|---|---|
| lens | |
| retina | |
| iris | |
| optic nerve | |
| ciliary muscles | |

3 Copy and complete the sentences. Use words from the list.

**cornea** **focused** **optic nerve** **pupil** **retina**

Light from an object enters the eye through the ____________. It then passes through the lens, where it is ____________ onto the _________ where there are light receptors. Electrical impulses are sent to the brain along the ______________.

The amount of light entering the eye is controlled by the size of the ___________.

4 Boxers receive many blows to the head. One effect of this can be that the retina becomes detached from the back of the eye. Someone with a detached retina cannot see clearly. Suggest a reason for this.

## Key ideas

- *Light enters the eye through the cornea.*
- *The cornea and the lens produce an image on the retina.*
- *The iris alters the size of the pupil in order to control the amount of light entering the eye.*

# 1:11 Drugs

Drugs are often in the news or are talked about by people. It seems that people have different attitudes towards drugs.

There are thousands of different drugs in the world; some are of benefit whereas others have caused lots of problems.

**a What do you think the people in the pictures mean when they talk about 'drugs'?**

## What is a drug?

Drugs are chemicals that affect the way your body works. Some drugs may help you when you are ill. Others may change the way you think, or how you feel pain. Some drugs are illegal whereas others can be bought in your local shop.

**b Make a table with three columns. From the list below group all the products containing drugs that are illegal, those that can only be bought by adults or with a prescription, and those that anyone can buy.**

| | | | |
|---|---|---|---|
| **alcohol** | **antibiotics** | **aspirin** | **cannabis** |
| **cigarettes** | **coffee** | **cola** | **cough sweets** |
| **heroin** | **tea** | | |

The table gives information about different types of drug.

Although paracetamol is safe to use, it can be dangerous if too much is taken. Excessive paracetamol leads to liver damage and death.

| Type of drug | Examples | Effects |
|---|---|---|
| stimulants | caffeine (found in tea, coffee and some soft drinks), ecstasy, nicotine, amphetamines | speed up the brain and make you feel more alert |
| sedatives | barbiturates | slow down the brain and make you feel sleepy; in excess can lead to unconsciousness and death |
| hallucinogens | LSD | may cause you to see or imagine things that are not really there (hallucinations) |
| painkillers | aspirin, paracetamol, morphine, heroin | suppress the parts of the brain that give the feeling of pain |

Heroin and morphine are given to terminally ill patients to relieve pain.

**c Which type of drug should be used for the following:**

**i a motorist who is feeling tired but still has some distance to drive;**

**ii a child who is suffering from toothache;**

**iii a man who wants to have a good night's sleep?**

**Did you know?**

Many road accidents happen because drivers are affected by drugs. Sometimes they are illegal drugs and sometimes they are prescription medicines. You should always read the information with prescription drugs to find out about the possible side-effects.

All drugs change the chemical reactions in our bodies. If your body gets used to the change it may become **dependent** on the drug. Your whole body's chemistry may have changed. When this happens you have become **addicted** to the drug.

If a person is addicted to a drug, taking the drug away can cause withdrawal **symptoms**. These can include dizziness, vomiting and hallucinations. As the body gets used to a drug, larger doses are required to have the same effect.

Sniffing glue can be fatal.

## Why do people take drugs?

People take drugs for lots of different reasons. If your body is not carrying out its normal chemical processes, a doctor may advise you to take a particular drug to help.

Taking some drugs can affect your performance, for example it may be dangerous for you to drive a car.

Other drugs are so dangerous that they are illegal in Britain. Heroin is highly addictive and causes lots of health problems. Cannabis is another illegal drug and can cause lung cancer.

Another problem can occur when drug takers inject drugs and then exchange needles, rather than using sterile ones. It is possible for viruses, particularly hepatitis and HIV, to be transferrerd from one person to another in this way.

**Did you know?**

It is estimated that half a million people in the UK are infected with the hepatitis C virus. Many of these people contracted the infection from sharing needles when injecting drugs, or because they received infected blood products in hospital.

About 125 young people die each year in Britain because of damage caused by solvents.

## Solvents

Some people sniff solvents to get a 'high'. These solvents include glue, lighter fuel, petrol and aerosols. For some people the effects of solvents are similar to alcohol. Others get visions and hallucinations.

Sniffing solvents is both dangerous and addictive. Death may result immediately from heart failure. Solvents damage the lungs and nerve cells. They damage the links between one nerve cell and another in the nervous system.

## Questions

**1** Complete the sentences. Use words from the list.

| | | |
|---|---|---|
| brain | chemical reactions | dependent |
| liver | reaction times | withdrawal symptoms |

Drugs change some ________________ in our bodies. As a result of these changes, we become ________________on the drug. If we stop taking the drug we will experience ________________ .

Solvents cause damage to our ________________ and________________ . Drugs can slow down our ________________ when we are driving.

**2** Aspirin was first discovered in the bark of willow trees. Now it is made from chemicals. Suggest why it is better to use aspirin tablets than aspirin extracted from willow bark.

**3** Design a leaflet to help people your age make the best decisions about drugs. You should explain how drugs can harm or kill you.

## Key ideas

- *Drugs change the chemical processes in your body.*
- *Drugs may become addictive and can lead to withdrawal symptoms if they are not taken.*
- *There are different types of drug.*

# 1:12 Alcohol and tobacco

## Cigarettes

Cigarettes are made from dried tobacco leaves.

Tobacco smoke contains a mixture of hundreds of different chemicals. Many of these chemicals are dangerous to the human body. Some chemicals stimulate cells to divide rapidly. This can lead to cancer. Tobacco affects the lungs in other ways as well.

### Tobacco affects breathing

Some chemicals in tobacco smoke paralyse the tiny hairs called cilia and cause the mucus glands to produce more mucus than they should. This disease is called bronchitis. After a while, mucus builds up in the lungs. This makes it hard to breathe and causes a smoker's cough. Coughing can cause further damage to the lungs.

Sometimes a smoker's bronchioles (the small tubes in the lungs) will become narrowed. This makes it harder to breathe and the smoker may become short of breath.

**Did you know?**
Smokers are about 30 times more likely to get lung cancer than are non-smokers. About one-third of deaths from cancer among men are caused by smoking. In spite of this, young people, particularly girls, are increasingly taking up smoking.

### Tobacco affects the circulation of blood

One of the poisonous substances in tobacco smoke is carbon monoxide. This prevents the red blood cells from carrying oxygen – so a smoker's blood carries less oxygen than does a non-smoker's blood. This is one reason why babies born to women who smoke are, on average, smaller at birth than babies born to non-smokers.

Some of the chemicals in tobacco smoke cause diseases of the heart and blood vessels. Many smokers die of heart attacks because the arteries that supply blood to the heart muscles become blocked.

## Alcohol

Alcohol is the oldest known drug. It affects the nervous system. Small amounts cause excitement and reduce inhibitions. But even small amounts slow down the body's reactions. That is why it illegal to drive a car when you have even small amounts of alcohol in your blood.

Larger amounts of alcohol lead to lack of self-control. One example of this is the fights that often break out when people drink too much.

Drinking too much alcohol can cause a person to lose consciousness and even go into a coma. Somebody who is unconscious risks choking to death on their own vomit. Every year we hear tragic stories of someone who has died after drinking too much alcohol.

**The amount of alcohol in the bloodstream can be worked out from this simple test.**

Drinking too much alcohol will lead to addiction. Many alcohol addicts die prematurely because of damage to their liver or brain.

It is important that people understand that they are likely to do damage to themselves by drinking alcohol. The government recommends that an adult man should not drink any more than four units (about two pints of beer) of alcohol per day.

The diagram shows the amounts of different drinks that contain one unit of alcohol.

**Each of these contains 1 unit of alcohol.**

**a John drinks a pint of beer, two glasses of wine and a single measure of whisky. How many units of alcohol has he drunk?**

**b Is this more than the government recommends?**

Alcohol is removed from the body at the rate of about one unit each hour.

Alcohol has a more dangerous effect on people under 18 years of age, so it is illegal for them to buy alcohol in Britain.

## Questions

1 Copy and complete the sentences. Use words from the list.

| | | | |
|---|---|---|---|
| **alveoli** | **brain** | **bronchitis** | **carbon monoxide** |
| **frequently** | **liver** | **lung** | **mucus** |
| **nervous** | **nicotine** | **reactions** | |

The addictive substance in tobacco is called ________ . Cancer results when cells divide too ________.

The disease that results from paralysis of the tiny hairs in the lungs is called ________ . The cells of the lungs of a person with this disease produce too much________ .

The gas in cigarette smoke that affects red blood cells is ________ .

Alcohol affects the ________ system, slowing down our ________ . Drinking too much alcohol can damage our ________ and ________.

2 Rick goes to a party and drinks heavily until 3 a.m. He gets a taxi home. Next morning he drives himself to work at 7 a.m. Why may this be unwise?

3 Tim drinks four pints of beer and two measures of whisky.

a How many units of alcohol has he drunk?

b Approximately how long will it take his body to break down the alcohol?

c Why is it impossible to predict exactly how long it would be before the concentration of alcohol in his blood was below the legal limit for driving?

## Key ideas

- *Tobacco smoke contains substances that cause diseases of the respiratory and circulatory systems.*
- *Alcohol affects the nervous system by slowing down reactions. It damages the liver and the brain.*

# 1:13 Keeping a balance

## Homeostasis

Tim buys an aquarium for tropical fish. He wishes to keep the temperature at exactly 28 °C. In the aquarium there is a thermometer, a heater and a thermostat.

He fills the aquarium with water at 20 °C. The heater switches on and the temperature of the water rises. When the temperature of the water reaches 28 °C, the thermostat turns off the heater. When the temperature starts to fall, the thermostat turns the heater back on. Only when Tim is sure the temperature is constant at 28 °C is it safe to put in the fish.

**a Which piece of equipment:**

**i measures the temperature in the aquarium;**
**ii heats up the water;**
**iii turns the heater on and off?**

Having steady conditions is called **homeostasis**. It is important that constant conditions are maintained in the body. There are many ways to do this. The following table summarises some of these.

| What needs changing | Organ responsible | How change is made |
|---|---|---|
| urea in the blood | kidney | urea excreted in the urine |
| blood too dilute | kidney | excess water excreted in the urine |
| blood too salty | kidney | salt excreted in the urine |
| too much carbon dioxide in the blood | lungs | breathing increases to excrete carbon dioxide into the air |
| not enough oxygen in the blood | lungs | breathing increases to absorb more oxygen from the air |
| too much glucose in the blood | liver | removes glucose from the blood and stores it as insoluble glycogen |
| too little glucose in the blood | liver | releases glucose from its store of glycogen |
| body temperature too high | skin | extra warm blood goes to the skin surface to lose heat<br>sweating increases |

These processes take place automatically. We do not need to think about them.

Many of these processes are controlled by **hormones**.

## Hormones

Hormones are chemicals. They are made in various organs in the body and are transported around the body in the bloodstream. Each hormone affects only particular organs. The organs affected by a hormone are called **target organs**. The table on the right gives some important hormones and where they are produced.

| Hormone | Produced by |
|---|---|
| adrenaline | adrenal gland |
| insulin | pancreas |
| oestrogen | ovary (female) |
| testosterone | testes (male) |

## Controlling body temperature

Just as Tim needs to keep the temperature of his aquarium constant, we have to keep our body at a constant temperature. One of the first tests a doctor does on a patient is to check the body temperature. A healthy person has a body temperature of about 37 °C.

In some illnesses a fever develops and the body temperature rises. If the body temperature rises by more than 4 °C, then doctors must take action to bring it down before permanent damage occurs. Proteins are **denatured** by high temperatures.

**b Suggest what happens to a person whose body temperature:**

**i gets too low;**
**ii gets too high.**

Body temperature is both monitored and controlled by the **thermoregulatory centre** in the brain. The table summarises ways in which the body keeps temperature constant.

| When the body is too hot | When the body is too cold |
|---|---|
| Hairs lie flat and no air is trapped. | Erector muscles in the skin cause hairs to stand erect, trapping air against the skin. This insulates the skin. |
| Sweat glands produce sweat, which consists of water and salts. When sweat evaporates from the skin's surface the skin is cooled. | Sweating is minimal. |
| People relax and avoid activity that will generate heat energy. | **Shivering** starts. During shivering the muscles contract, increasing the rate of respiration and therefore producing heat. Blood flowing through the muscles is warmed. |

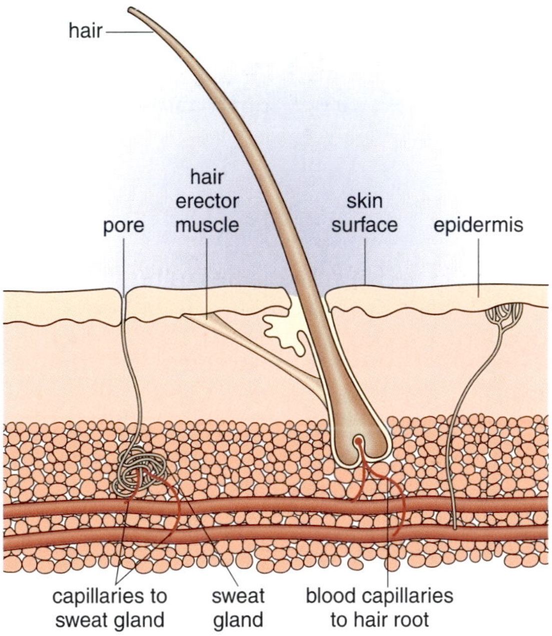

**A cross-section through the skin to show the parts involved in temperature regulation.**

## Questions

**1 List the different ways in which your body controls its temperature.**

**2 Look at the diagram of the cross-section of the skin.**

**a Suggest the role of the outer layer of the skin (epidermis).**

**b What can the skin do to conserve or lose heat energy?**

### Key ideas

- *Homeostasis is keeping internal conditions in the body constant.*
- *A hormone is a chemical produced by one part of an organism to control a process in another part of the organism.*
- *Hormones are transported in the bloodstream to the target organ.*
- *Shivering results in an increased rate of respiration, which warms blood flowing through the muscles.*

# 1:14 Getting rid of waste

## Urea

The body needs amino acids to build new cells. We get these amino acids by digesting the protein in our food.

The body cannot store amino acids, so it has to get rid of excess – the ones it cannot use immediately. Amino acids contain energy. Enzymes in the liver split the excess amino acids into two parts: the acid part is used in respiration, and the amino part is made into urea.

Urea is a poisonous waste product so the body has to get rid of it.

1 Urea is transported by the blood to the **kidneys**.
2 The kidneys extract urea, excess water and excess ions from the blood and make **urine**.
3 Urine flows from the kidneys to the **bladder**.
4 Urine is stored in the bladder until we release it when we urinate.

Urine consists of:

- Urea;
- Excess water (water the body does not need);
- Ions that the body does not need.

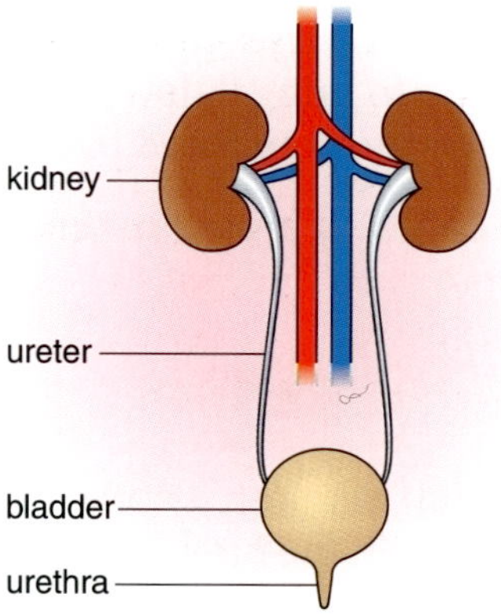

**The urine leaves the kidneys via the ureter and is stored in the bladder.**

## The kidneys

The kidneys regulate the amount of water in the body. The diagram shows a cross-section of a kidney.

The diagram opposite shows the amounts of water entering and leaving the body in a day.

**a What do you notice about the volumes of water entering and leaving the body each day?**

**b Why is it important to drink lots of liquids when you are suffering from diarrhoea?**

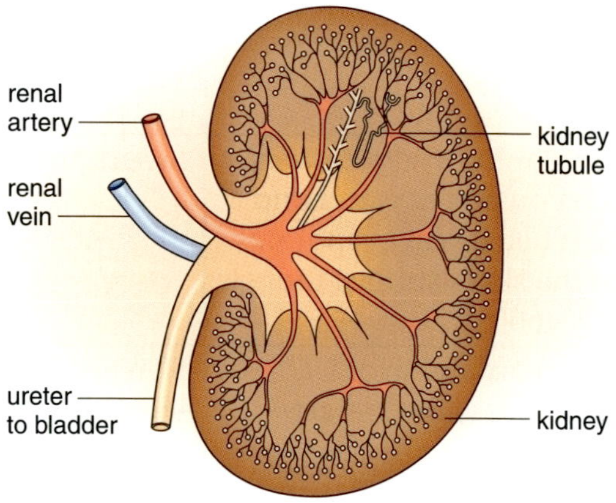

**A cross-section of the kidney showing the internal structure.**

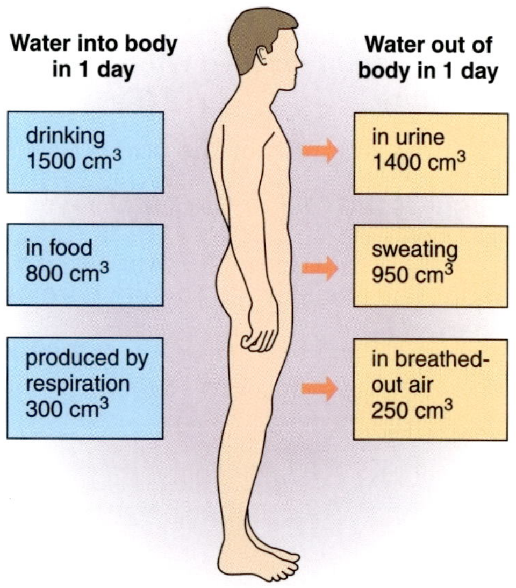


**Did you know?**

Each kidney can filter 1.3 litres of blood each minute.

If we were unable to re-use the water in our bodies we would spend all day on the toilet and drinking water.

## When kidneys fail

If the kidneys fail then waste materials build up in the body and begin to poison it.

There are two methods to treat people with kidney failure. The best method is to remove the failed kidneys and replace them with a healthy kidney. The healthy kidney can be transplanted from a person who has just died. Unfortunately, however, there is a big shortage of kidneys available for transplants.

When kidneys are not available, patients are connected to a mechanical kidney called a **dialysis machine**. This machine cleans the blood in the same way that the kidneys do, removing urea and other waste materials and regulating the amount of water and ions in the blood.

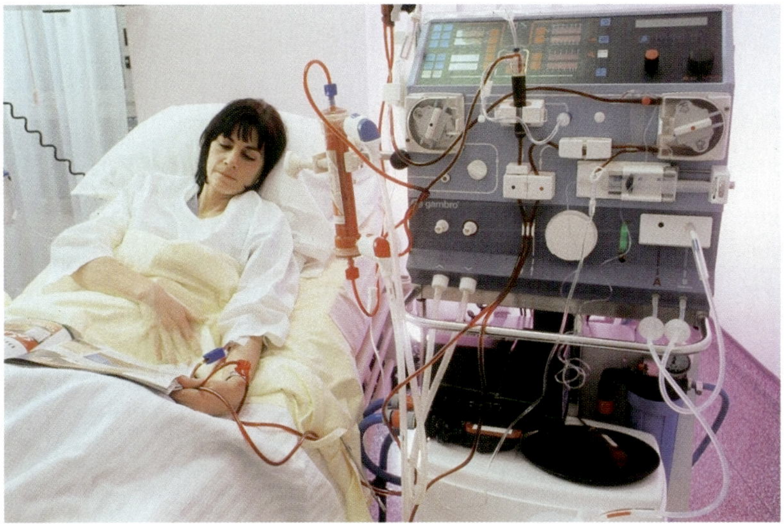

**This patient is being treated with a dialysis machine.**

## Questions

1 Write down the names of four substances filtered from the blood by the kidneys.

2 Copy and complete the following sentences. Use words form the list.

| | | |
|---|---|---|
| amino acids | bladder | ions |
| kidneys | liver | water |

Urea is formed by the breakdown of ______ in the ______. Urine is produced by the ______ and is stored in the ______. Urine consists of urea, excess ______ and excess ______.

3 Explain why you would produce a smaller volume of urine each day while in a tropical country than you do each day in Britain.

### Key ideas

- *The kidneys filter the blood and reabsorb all the dissolved ions, sugar and water needed by the body.*
- *The kidneys release urea, excess ions and water as urine.*
- *The kidneys keep the balance of water in the body constant.*

# End of module questions

## Low demand

**1** Oxygen is transported round the body by:

**A** plasma

**B** platelets

**C** red blood cells

**D** white blood cells

**2** John has a bacterial infection. His doctor gives him:

**A** antibodies

**B** antibiotics

**C** aspirin

**D** antacids

**3** Which of these drugs is a stimulant?

**A** barbiturates

**B** caffeine

**C** heroin

**D** paracetamol

**4** Drinking alcohol can:

**A** damage the liver

**B** damage the skin

**C** make you more alert

**D** make your reactions faster

**5** A network of fibrin strands is used to:

**A** fight infection

**B** filter the blood

**C** make a scab

**D** strengthen blood vessel walls

**6** Which of these is an illegal drug?

**A** aspirin

**B** caffeine

**C** heroin

**D** tobacco

The drawing shows the daily input and output of water for an adult.

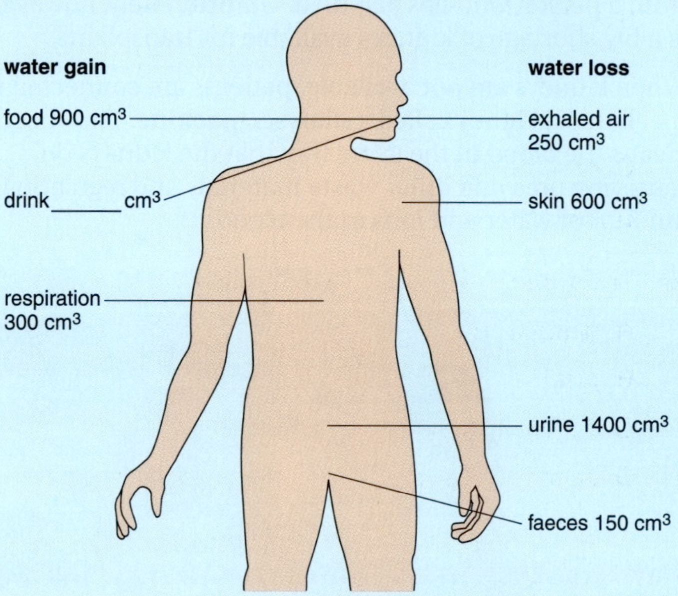


**7** Refer to the figure above. How much water did the person drink in one day?

**A** 400 $cm^3$

**B** 1000 $cm^3$

**C** 1200 $cm^3$

**D** 2400 $cm^3$

**8** Refer to the figure above. What fraction of the water was lost through the skin?

**A** $\frac{1}{4}$

**B** $\frac{1}{10}$

**C** $\frac{1}{40}$

**D** $\frac{1}{400}$

**9** Tom produces urine in the:

**A** bladder

**B** kidneys

**C** penis

**D** testes

**10** Normal human body temperature is:

**A** 17 °C

**B** 27 °C

**C** 37 °C

**D** 47 °C

**11** What do red blood cells transport around the body?

**A** energy

**B** food

**C** hormones

**D** oxygen

**12** The liquid in which all blood cells float is:

**A** gastric juice

**B** plasma

**C** sugar solution

**D** water

## Standard demand

**13** Where in the eye are the light receptors?

**A** the iris

**B** the lens

**C** the optic nerve

**D** the retina

**14** How many of the receptor types in the list are found in the skin?

**chemical** **light** **temperature** **touch**

**A** 1

**B** 2

**C** 3

**D** 4

**15** White blood cells fight infection by producing:

**A** antibiotics

**B** antigens

**C** antiseptics

**D** antibodies

**16** Meningitis bacteria produce harmful chemicals that cause septicaemia. These chemicals are:

**A** antibodies

**B** antibiotics

**C** antitoxins

**D** toxins

**17** Platelets:

**A** fight infection

**B** help to heal wounds

**C** produce new red blood cells

**D** transport hormones around the body

**18** Which organ produces urine?

**A** the bladder

**B** the kidneys

**C** the liver

**D** the lungs

**19** Which organ stores urine?

**A** the bladder

**B** the kidneys

**C** the liver

**D** the lungs

**20** The diagram below shows an investigation to find out how food is absorbed from the digestive system.

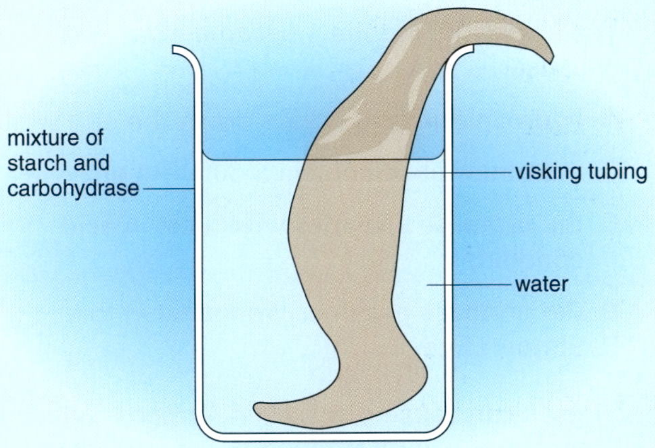


The visking tubing used in the investigation works in the same way as:

**A** the alveolar surface

**B** the wall of the small intestine

**C** the wall of the capillaries

**D** the cell membrane

21 Refer to the diagram in Question 20. After about 30 minutes the water around the tubing will contain:

**A** carbohydrase

**B** glucose

**C** protein

**D** starch

22 The diagram below shows an investigation carried out to find out about the enzymes produced by the stomach. When the egg white (a protein) is digested, the contents of the test tubes change from cloudy to clear.

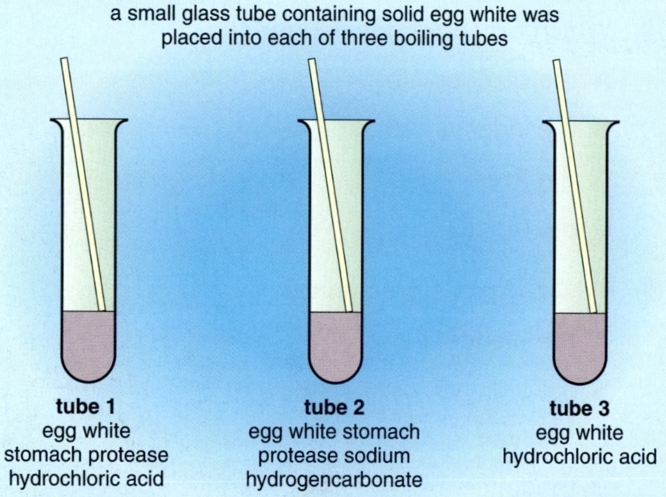


After 30 minutes the egg white in tube 1 became clear whereas the egg white in the other tubes remained cloudy.

The egg white in tube 1 became clear first because:

**A** hydrochloric acid breaks down the egg white

**B** the other tubes contain a different enzyme

**C** the protease enzymes work best in acid conditions

**D** the protease enzymes work best at this temperature

23 Viruses can cause diseases in humans. Which of the following parts is found in a virus?

**A** cell membrane

**B** cytoplasm

**C** nucleus

**D** protein coat

24 Which part of the digestive system contains villi?

**A** the large intestine

**B** the pancreas

**C** the small intestine

**D** the stomach

# Module 2 – Inheritance and survival

Why are humans different from other animals, and why are all humans different from each other? Scientists have taken hundreds of years to find the answers to these questions, and they are still working on them!

This module is divided into three parts.

1 In the first topic you will find out how characteristics such as sex and hair colour are inherited from your parents through the chromosomes that are passed on.

2 In the second topic you will learn how the information for different characteristics is carried on small parts of chromosomes called genes and how these can be changed artificially.

3 The third topic is about survival in different conditions and how the large population of humans on the planet is changing the environment.

Try these questions first.

1 What is the name of an animal male sex cell?
2 Whereabouts in the body are male sex cells produced?
3 What happens at fertilisation?
4 A sparrowhawk eats a wren. Which is the predator and which is the prey?
5 What three important things does a habitat provide?

# 2:1 There is no-one like you!

There are many different kinds of animals and plants. We call these **species**. You look very similar to other people around you because you belong to the same species.

Even though you look similar to other people you can still be distinguished from them. This is because of your **characteristics** or features such as hair colour, eye colour and height. These are just a few of the many characteristics that distinguish you from everyone else. There are millions of people in the world yet no two people are exactly the same. You are unique!

**a What important characteristics do the police use to identify criminals even if they have left the scene of a crime?**

Can you tell the difference between these sheep? Probably not – but other sheep can!

## Spot the difference

Young animals and plants often look like their parents. The puppy in the photograph has many similar characteristics to its parents.

**b Give two examples of characteristics that the puppy has in common with:**

**i all other dogs;**
**ii its parents.**

Young animals and plants look like their parents because some characteristics are **inherited**. These characteristics are passed on from parents to their offspring.

The puppy in the photograph has many similar characteristics to its parents.

## Your make up

Inside almost every cell in your body there are strands of a chemical called DNA. These strands:

- contain all the information that you inherited from your parents
- determine characteristics such as the natural colour of your eyes and hair
- determine your sex.

They do not determine all your characteristics, for example whether you have your ears pierced – some things are best left to individual choice.

The strands of DNA are in the nucleus of each cell. This is the part of the cell that controls its activity. When the cell divides to reproduce, the strands of DNA are copied. This means that the new cells all have the same information.

## Existing in pairs

The strands of DNA are called **chromosomes**. You inherit 23 **chromosomes** from each parent. This means that you have 46 chromosomes altogether. These chromosomes are present in most of your body cells, so you actually have millions of identical copies of them.

The number of chromosomes in each normal cell is called the **diploid number**. Diploid means two sets of chromosomes, one set from each of your parents. Different animals have different diploid numbers. The diploid number is a characteristic of the species. The diploid number in humans is 46.

The chromosomes inside the nucleus of each body cell occur as pairs. Each chromosome that you inherited from your father has a partner from your mother. The photograph shows the chromosome pairs for a female. The pairing is slightly different for a male.

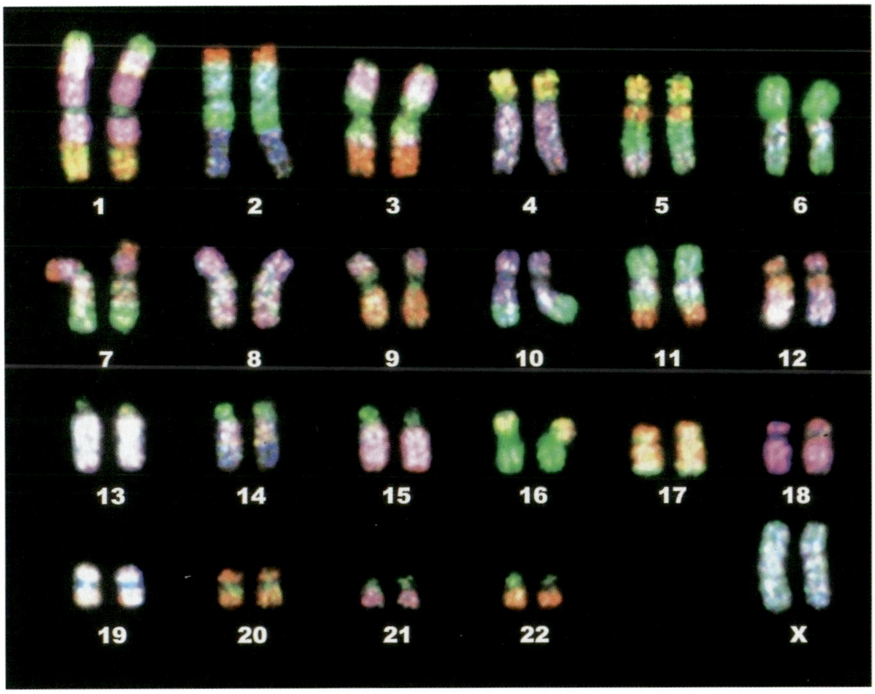


The 23 pairs of chromosomes in a human female. (The chromosomes have been set out in pairs here so that they can be examined. In cells the chromosomes are jumbled together.)

**c** Which of these are determined by your chromosomes?

eye colour hair colour hair length
sex skin colour

**Did you know?**
A child with Down's syndrome has an extra chromosome.

## Questions

1 Copy and complete the sentences. Use words from the list.

characteristic chromosome diploid species

A feature such as height is called a ______________.

Each strand of DNA in the nucleus of a cell is a ______________.

The number of chromosomes in most body cells is called the ______________ number.

2 a State five characteristics that are common to all humans.

b Explain how identical twins are different from non-identical twins.

c In what ways are all cats similar to each other, and in what ways are they different?

3 A horse has 66 chromosomes in a normal cell.

a What is the diploid number?

b How many chromosomes did it inherit from:

i its mother;
ii its father?

4 In which part of a cell are the chromosomes?

### Key ideas

- *Each normal human body cell has 46 chromosomes – the diploid number.*
- *The chromosomes occur as 23 pairs.*
- *One set of chromosomes comes from the mother and one set from the father.*

# 2:2 Some cells are different

Not all the body's cells have a nucleus. Mature red blood cells do not have nuclei, so they do not carry inherited information.

All your chromosomes came from your parents. When it comes to what you inherit, they have an equal contribution to the matter.

Chromosomes from your parents are passed on in the **gametes**. Gametes is another name for the sex cells.

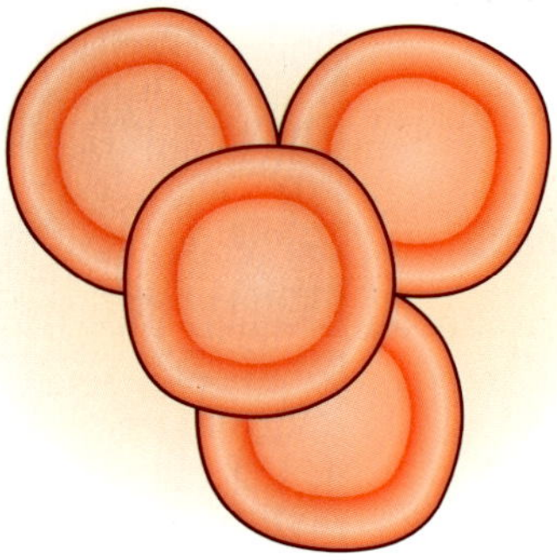

These red blood cells do not have a nucleus.

## Male gametes

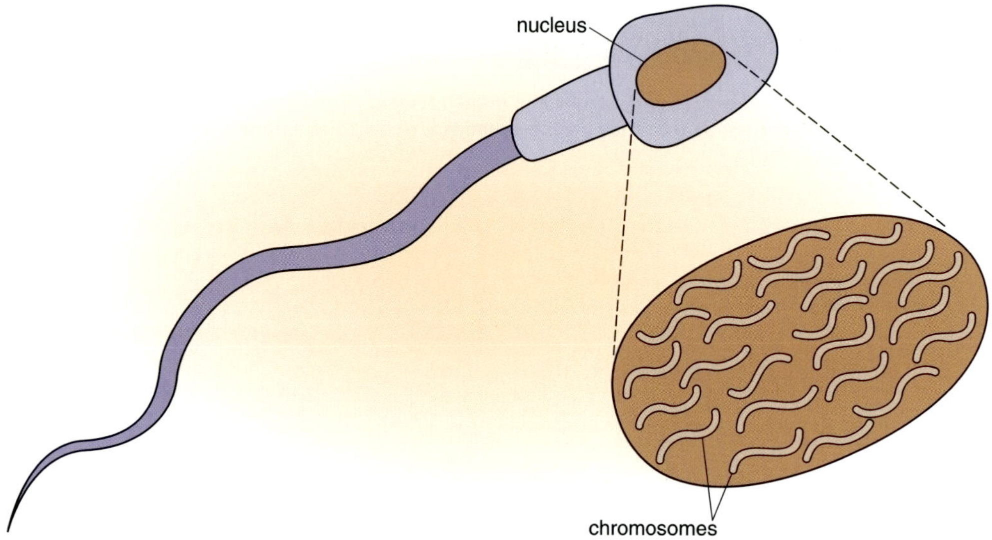


The nucleus of the sperm is in the head and contains 23 chromosomes.

Male gametes are called **sperm**. Sperm are made continuously in the **testes**. The nucleus is inside the head of the sperm. Inside this nucleus there are 23 chromosomes – one from each pair in a normal body cell. The number of chromosomes in a gamete is called the **haploid number**. Haploid means one set of chromosomes. Normal body cells have two sets (the diploid number).

**a What is the haploid number of a normal human cell?**

And here's another difference between normal body cells and gametes. In a normal body cell, which has 46 chromosomes, all the chromosomes are identical to those in other normal body cells. Not so with sperm. Because each chromosome can be either one of the two in a pair, there are more than 8 million different possible combinations of chromosomes.

> **Did you know?**
> Sperm need a constant supply of energy so that they can travel through the uterus. Before being released from the penis, they are mixed with a glucose solution to provide this energy.

## Female gametes

A female gamete is an egg, or **ovum** (plural **ova**). A newborn female has all the ova that she will ever need, and a lot more besides. She does not make new ova. The ova develop and mature in the **ovaries** during the menstrual cycle.

Like the sperm, the nucleus of an egg contains 23 chromosomes. And, also like the sperm, there are more than 8 million possible combinations of chromosomes.

An egg is much bigger than a sperm because it carries a food store as well as a nucleus.

**b** **What are the names of the male and female gametes?**

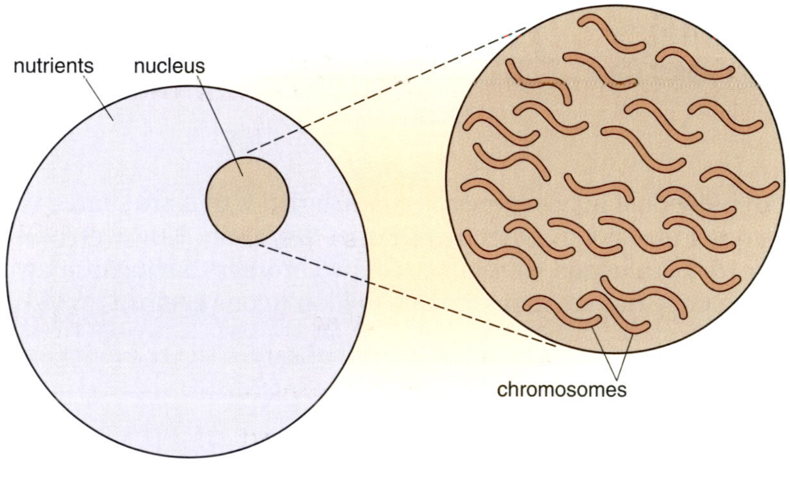


**The nucleus of the ovum contains 23 chromosomes.**

## Pairing up

Gametes contain single chromosomes whereas those in normal body cells are in pairs. A new individual is formed when the nuclei of a male and a female gamete fuse, or join together. The diagram shows how two cells combine to make one.

In this way, two haploid cells fuse to make a diploid cell.

**c** **Explain why a diploid cell has twice as many chromosomes as a haploid cell.**

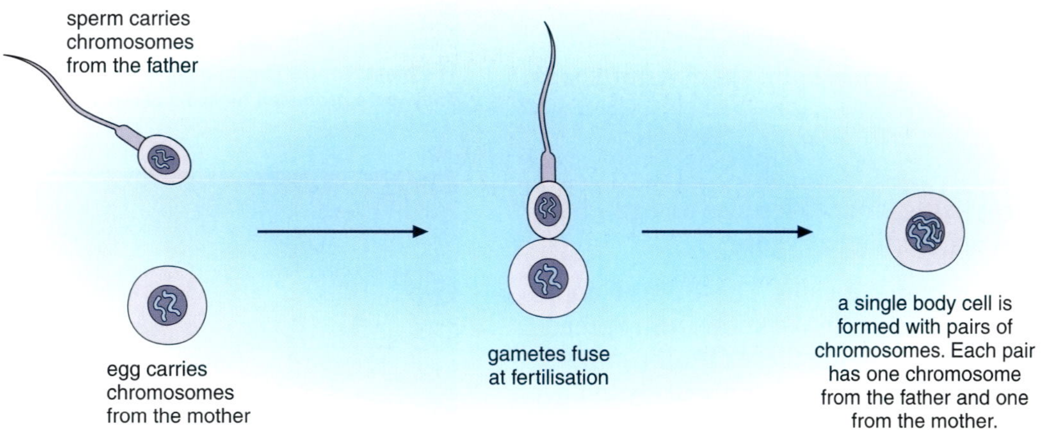


**Sexual reproduction produces variation because chromosomes separate when gametes are formed, and combine by chance when gametes fuse at fertilisation.**

## Questions

1. What type of cells are gametes?
2. How is the number of chromosomes in a gamete different from the number of chromosomes in a normal body cell?
3. How is the arrangement of chromosomes in a gamete different from that in a normal body cell?
4. In a housefly, there are 12 chromosomes in a diploid cell. How many chromosomes are there in a housefly haploid cell?

### Key ideas

- *Gametes are sex cells.*
- *Male sex cells are called sperm.*
- *Female sex cells are called eggs or ova.*
- *A gamete is a haploid cell – it contains only one set of chromosomes.*
- *A human gamete has 23 chromosomes.*

# 2:3 Variety of life

An individual may have some characteristics that are shared with a brother or sister or other members of the family. However, unless they are identical twins, the chances of two brothers having exactly the same characteristics are more than 64 million to one against!

**a What are the chances of a brother and a sister having exactly the same characteristics?**

## Asexual or sexual

Some species can reproduce without two parents being involved. For example, new geranium plants can be produced by taking cuttings from the parent plant – this is **asexual** (non-sexual) reproduction. Alternatively, pollen from the male part of a flower can fertilise eggs in the ovaries of another flower – this is **sexual** reproduction because it involves both male and female.

In asexual reproduction the offspring are identical to the parent because they have identical chromosomes. The offspring are called **clones.**

In sexual reproduction the offspring have some characteristics of each parent because half their chromosomes come from the male parent and half come from the female parent.

**b A gardener wants a flowerbed full of pink geraniums. Explain whether he should grow the plants from cuttings or from seeds.**

The plants in this greenhouse are all identical. They were produced by asexual reproduction.

## Different people

At **fertilisation**, millions of sperm battle it out to fertilise an egg – but only one succeeds. The fertilised egg is called a **zygote**.

The sperm carries 23 chromosomes in the nucleus. These are all copies from the father's set of 46 – one from each pair. We know from Spread 2:2 that there are millions of ways of combining these chromosomes. This means that two offspring from the same parents are likely to have some chromosomes in common with each other and some chromosomes that are different.

The chromosomes they have in common lead to shared characteristics. Different characteristics are the result of different chromosomes.

**c Look at the picture of the family. What similarities and differences can you see in the children?**

Members of a family show some similarities and some differences because of the chromosomes they inherit.

## Boy or girl?

For each chromosome in a sperm there is another matching chromosome in the egg. (There is one exception – the sex chromosomes – which will be explained later.) Each chromosome controls different characteristics. For example, one chromosome in the sperm carries information about eye colour, as does its partner chromosome from the egg. But the specific information that they carry about eye colour may be different. We come back to this in Spread 2:7.

One pair of chromosomes, called the **sex chromosomes**, determines sex. Females have two sex chromosomes that are the same shape. These are described as XX. Males have one X chromosome and one that is a different shape. These sex chromosomes are described as XY.

- All ova contain only an X sex chromosome.
- Some sperms contain an X sex chromosome and some contain a Y sex chromosome.

This means that your sex was decided by the sex chromosome you inherited from your father. If you are female then your sex chromosomes are XX and you must have inherited an X chromosome from your father. If you are male then your sex chromosomes are XY and you must have inherited a Y chromosome from your father.

**Did you know?**
When a mother-to-be has an ultrasound scan, this reveals the sex of the fetus. This means that the parents can know the sex of their baby before it is born, although many prefer not to know.

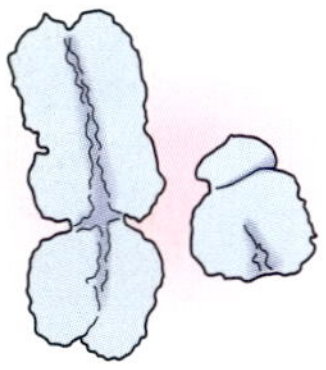

**X and Y chromosomes get their names from their shape.**

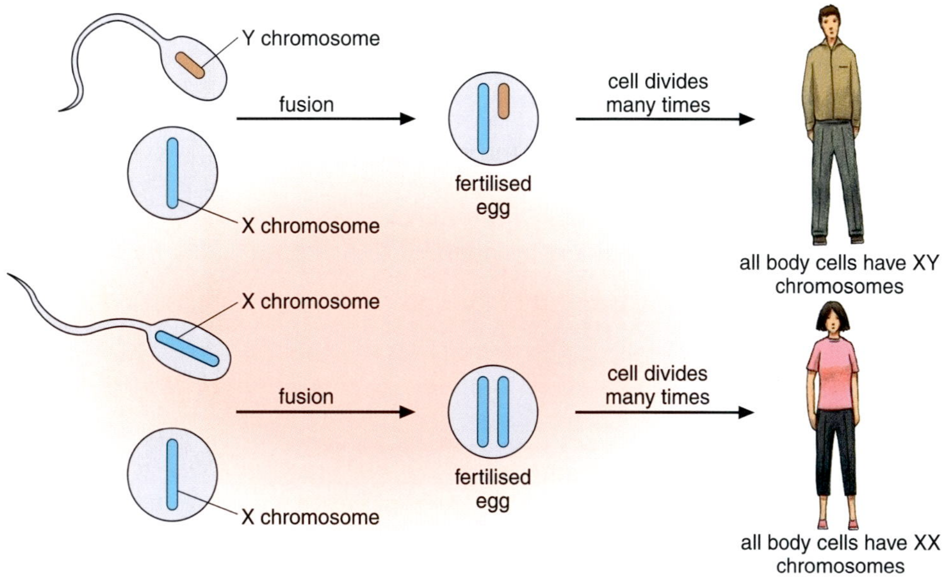


**Determination of sex. If the sperm contains an X sex chromosome, the fertilised egg will become a female. If the sperm contains a Y sex chromosome, the fertilised egg will become a male.**

**d What are the chances of a baby being female? Explain your answer.**

## Questions

1 How many parents are involved in asexual reproduction?

2 Explain why the offspring from asexual reproduction are all identical.

3 a Copy and complete the diagram to show how the sex chromosomes from a mother and a father can pair up.

b How many of the possible outcomes would result in a girl?

c How many would result in a boy?

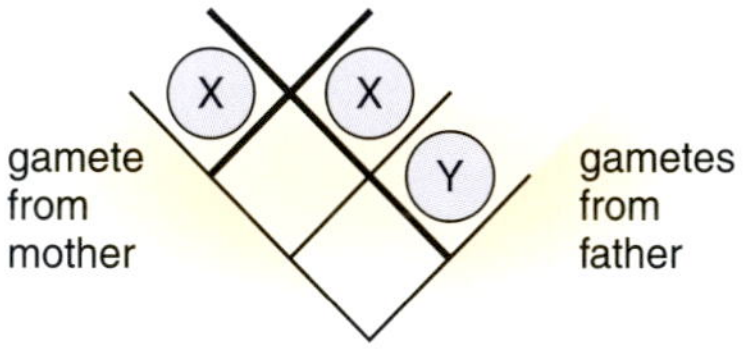


4 A brother and a sister have the same parents.

a Explain why they share some characteristics.

b Explain why they have some different characteristics.

## Key ideas

- *In sexual reproduction, a new individual inherits one set of chromosomes from the mother and one set from the father.*
- *The result is that the offspring are different from the parents but share some characteristics.*
- *Sex is determined by the sex chromosomes. Females are XX and males are XY.*

# 2:4 Hormones

Hormones are chemicals that are produced in organs called **glands**. Hormones are responsible for changes in the body. For example, some hormones control growth, others are responsible for maintaining the water level in the blood. The production of sperm and the release of eggs are processes that are controlled by the **sex hormones**.

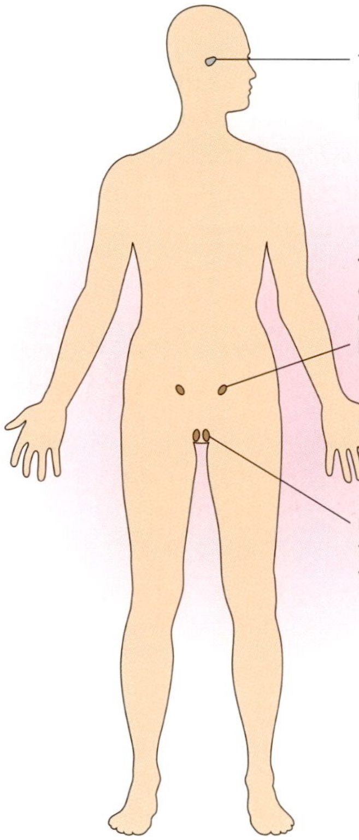


## Male hormones

The male sex hormone is called **testosterone**. It is produced in the testes. During puberty the production of testosterone increases. This has the following effects.

- The testes produce sperm – this continues for most of the man's life.
- The secondary sexual characteristics develop – the voice deepens, body hair grows and the shoulders broaden.

**a Which hormone is responsible for the growth of hair on the face?**

## Female hormones

The female sex hormones are called **oestrogen** and **progesterone**. They are produced by the ovaries.

Oestrogen controls the development of secondary sexual characteristics, including:

- growth of body hair
- development of breasts (mammary glands)
- broadening of the hips
- onset of **menstruation**.

### The menstrual cycle

Both oestrogen and progesterone are involved in controlling the menstrual cycle. This is the monthly cycle in which an egg is released and the uterus is made ready to receive an embryo.

The **menstrual cycle** has various stages.

1 During menstruation the old lining of the uterus falls away and passes out through the vagina.

2 During the early part of the cycle, oestrogen is released. This causes the repair and thickening of the lining of the uterus. It also causes an ovum to ripen in the ovaries.

3 Ovulation occurs midway through the cycle. A mature egg is released from an ovary. This is controlled by a hormone from the pituitary gland.

4 In the second half of the cycle progesterone is released. This causes more thickening of the lining of the uterus and prevents it from falling away. If the egg is not fertilised the progesterone production stops and menstruation occurs.

**Did you know?**
Anabolic steroids have a similar effect to testosterone. They can be used to increase muscle growth artificially. This would give an athlete more strength and an unfair advantage against other athletes. This is why the use of anabolic steroids is banned by sporting organisations.

The diagram shows the changes in the uterus and in the hormone levels during menstruation.

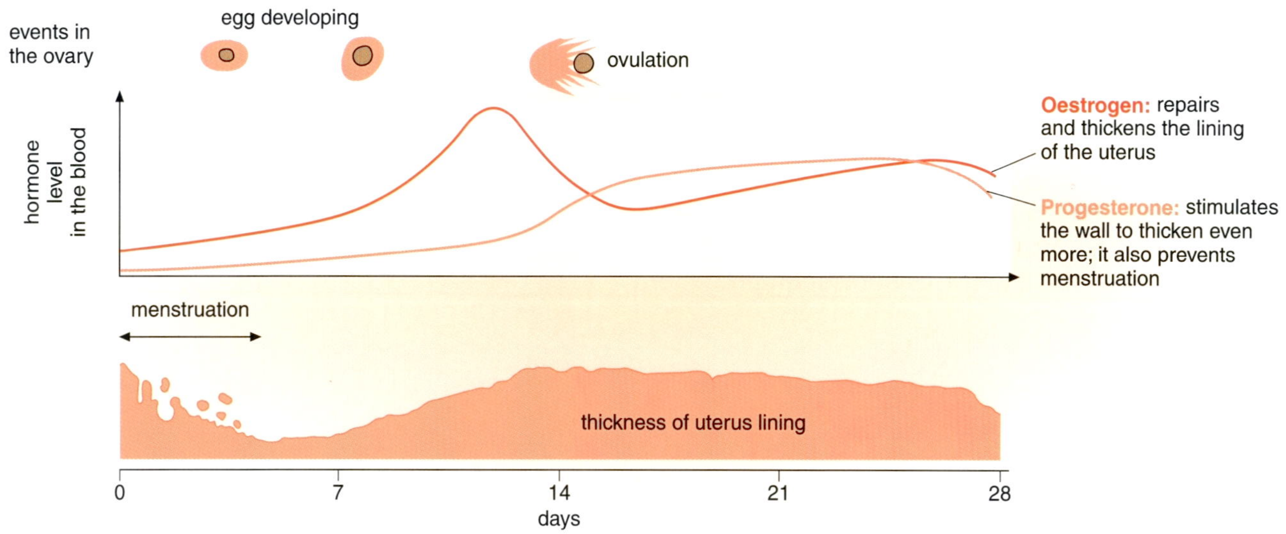


**Hormonal control of the menstrual cycle.**

**b** **Why is it important that the production of progesterone is maintained if the egg is fertilised?**

**c** **On which day of the menstrual cycle is the egg released?**

## Treating infertility

A couple may be unable to have a child for many different reasons. Some causes of infertility can be treated with manufactured hormones.

- The woman can receive injections of the hormone that causes ovulation. This sometimes causes several eggs to be released and can lead to multiple births.
- Oestrogen and progesterone can be used to control the menstrual cycle so that the lining of the uterus thickens ready to receive an embryo.

## Questions

**1** **a** State whether each of these hormones is a male sex hormone or a female sex hormone.

  **i** oestrogen;
  **ii** progesterone;
  **iii** testosterone.

**b** Where is each of the hormones listed in part **a** produced?

**2** At which part of the menstrual cycle does:

**a** the lining of the uterus fall away;

**b** ovulation occur;

**c** the production of progesterone stop?

**3** Why is the menstrual cycle called a *cycle*?

### Key ideas

- *Hormones cause changes to take place in the body.*
- *Testosterone is a male sex hormone.*
- *Oestrogen and progesterone are female sex hormones. They control the menstrual cycle.*
- *Some causes of female infertility can be treated with manufactured sex hormones.*

# 2:5 The double helix

All the information that is inherited from your parents is passed on through the chromosomes. The human body is a complex machine, so there is a lot of information to be passed on. There are instructions for growth and development, as well as the characteristics of the individual. Each chromosome carries hundreds of pieces of information. This information is written in a special code.

This is just a small section of a chromosome.

## Chromosomes and genes

Each piece of information that you inherit is carried on a **gene**. This is a small section of a chromosome. A single gene controls just one characteristic. However, one characteristic can be controlled by more than one gene. There are genes for the colour of your eyes, and genes for the colour of your hair. There are also genes that influence your height and weight.

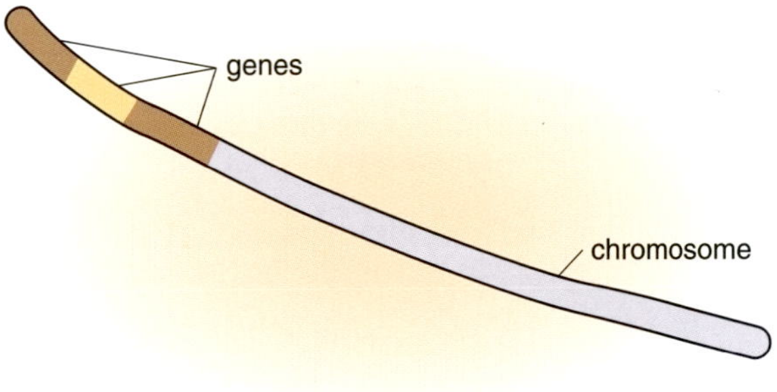


Each section of chromosome is a different gene.

The chromosomes in the nuclei of your cells are in pairs. Each one in the pair carries genes that control the same characteristics. For most characteristics, you have two genes, one from each parent. Sex chromosomes in males are the exception – the X and Y chromosomes have a different shape and carry some different genes

**a Explain why the sex chromosomes in females are the same shape.**

## The double helix

Genes are sections of a long chain molecule called **DNA**. Until 1950 no-one was really sure if DNA was actually the material involved in inheritance.

Two teams led the race to discover the secrets of DNA. Maurice Wilkins and Rosalind Franklin used X-rays to discover the shape of DNA. Meanwhile, James Watson and Francis Crick were building a three-dimensional model of the molecule. Using the results of Franklin's X-ray experiments, Crick and Watson were able to complete their model and won the Nobel Prize for their work. Sadly, Franklin died of cancer before the prize was awarded.

Watson and Crick's model showed that a DNA molecule is made of two very long **strands**. The strands are twisted into a shape called a **double helix**. It looks like a twisted ladder.

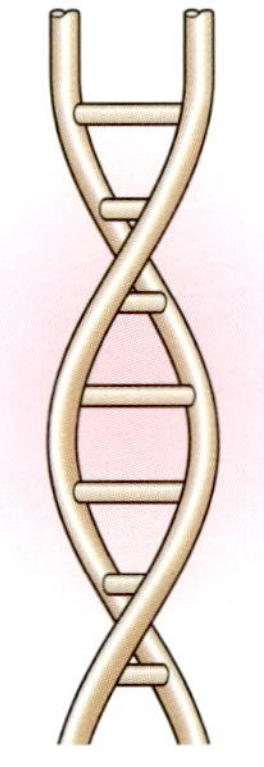

The double helix structure of DNA.

The rungs of the ladder join the two strands together. Each rung is madc of a pair of chemicals called **bases**. Because of their structure, the bases can only join together in a certain way:

- Adenine (A) joins with thymine (T).
- Cytosine (C) joins with guanine (G).

These bases carry information in a code.
The code is the order or sequence of the bases. Different orders of bases represent different information.

**b Explain why adenine cannot pair up with guanine.**

As well as carrying all the genetic instructions, DNA has to be able to copy itself. This is known as **replication** and takes place before a nucleus splits in two when a cell divides. In this way each new cell has an exact copy of the genes and chromosomes.

**c Which four chemicals are used in the genetic code?**

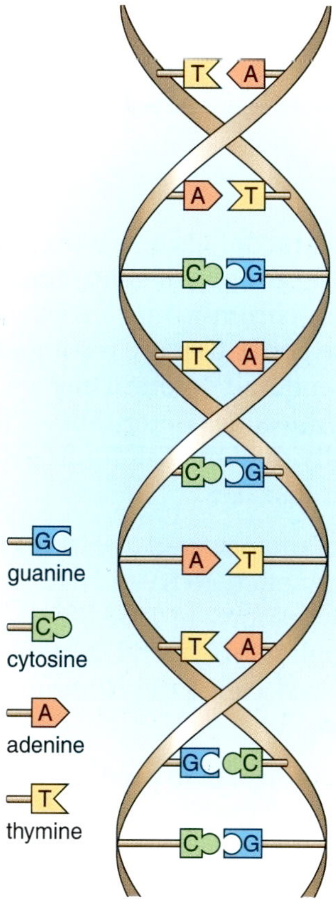

**Joining of the strands in DNA.**

## Questions

1 Copy and complete the sentences. Use words from the list.

chromosome DNA gene

Each piece of inherited information is in the form of a ______________. Hundreds of genes are strung together to make a ______________. Chromosomes are made up of a chemical called ______________.

2 a What is the name of the chemicals that link the DNA strands together?

b Copy and complete the sentences.

Adenine (A) joins with ______________ (T).

Cytosine (C) joins with ______________(G).

3 a Why does DNA need to copy itself?

b What is the process called?

4 a Explain the difference between a gene and a chromosome.

b How many genes are needed to control a characteristic such as eye colour?

c How many genes do humans have for eye colour?

d Whereabouts in the body are the genes for eye colour?

**Did you know?**
If the DNA from one human cell were unravelled, it would be a quarter of a million miles long.

## Key ideas

- *Each gene carries one piece of inherited information.*
- *Genes are sections of chromosomes.*
- *DNA is the chemical that carries the coded messages about inherited information.*
- *The structure of DNA is a double helix.*

# 2:6 Mapping the genes

Scientists know the details about only a handful of the thousands of genes that make up human chromosomes. They know about the genes that are responsible for eye colour and hair colour. They suspect that some genes determine whether an individual may suffer from cancer or other diseases.

The Human Genome Project will provide a huge amount of information about genes and their functions

There are lots of unanswered questions in the science of genetics. To try to answer some of these questions, in 1986 life scientists throughout the world started a very ambitious task called the ***Human Genome Project***.

The aim of this project is to produce a complete map of all the genes on the 23 pairs of human chromosomes. The first draft of this map, showing the positions of all the genes, was published in 2001. The next stage will be to find out what each gene actually does.

**a What is the aim of the Human Genome Project?**

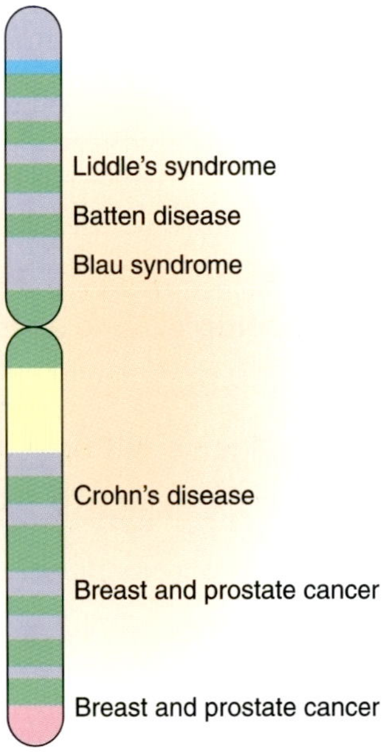


Some disorders that correspond to particular genes on human chromosome 16.

## Why map the human genome?

There are lots of potential benefits to people. These include:

- Health – it may be possible to find out whether someone is likely to be affected by a particular illness. If so, then the individual can be treated with drugs to reduce the risk.
- Lifestyle – people can be advised about the best diet to keep them healthy and which food chemicals could be harmful.
- Drugs – these can be developed so that they target particular genes. People with the same disorder could be given a different tailor-made drug to treat it.
- Gene therapy – this will enable scientists to replace genes that have errors resulting from mutations or errors in copying.

**b Explain how the results of the Human Genome Project could lead to an improvement in people's health.**

**Did you know?**
It is estimated that each human has 100 000 genes.

## What are the dangers?

Within a few years, scientists may have identified the genes that are responsible for causing certain cancers. A sample of DNA taken from a person could show how likely it is that they will develop a cancer in the future. This raises certain questions:

- Should an individual be told that he or she has an increased risk of developing cancer?
- Should the information be available to employers?
- Should insurance companies have a right to genetic information about people they insure?

Telling a person that he or she has a high risk of developing cancer in the future would certainly affect his or her life. It could cause them to be anxious or depressed and might even lead to suicide. However, it might encourage them to live life to the full, and to take care of their health to help reduce the risk of the cancer developing.

If the information were more widely available, it could affect a person's employment prospects. Life insurance companies are often unwilling to insure people who have a high chance of suffering from cancer.

**c Would you want to be told that you had a high risk of developing cancer? Give arguments for and against being told.**

## Footing the bill

The cost of the Human Genome Project is immense. Contributors to the cost include governments, charities and private and public companies. Manufacturers of drugs and medicines have a lot to gain from knowledge of the human genes because it will enable them to develop new drugs that are more effective at treating certain types of illness. This is why they are willing to invest a lot of money in the Human Genome Project.

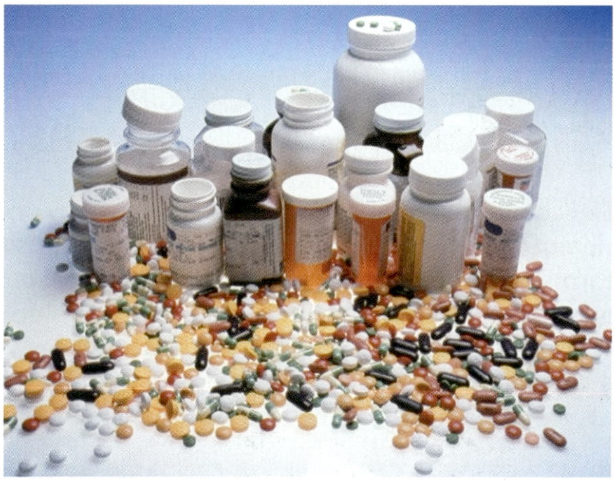

Knowledge of the human genome will help drug companies to develop new ways of treating inherited disorders such as cystic fibrosis.

## Questions

1 **Why is the Human Genome Project taking a long time to complete?**

2 **What are the benefits of scientists from several countries working on the Human Genome Project?**

3 **Explain why it may not be a good idea to tell a person that they have a high risk of contracting a fatal illness.**

### Key ideas

- *The aim of the Human Genome Project is to produce a map that shows the position of every gene on the human chromosomes.*
- *The results of the Human Genome Project will help medical science to develop new ways of treating disorders.*
- *The results could be used in ways that are not in the best interests of individuals.*

# 2:7 Genes in different forms

These people both carry genes for eye colour, but they have different alleles.

Eye colour in humans is determined mainly by one gene. People do not all have the same eye colour, so there must be different forms of the gene which carry different codes. Some forms of the gene have the code for brown eyes, some for blue eyes and so on. These different forms of a gene are called **alleles**. A child inherits a gene for eye colour from each parent, but they may be different alleles carrying different codes and giving different messages.

**a In what way can alleles of the same gene be different?**

## Who decides?

For most characteristics, an individual inherits one gene from each parent. If both alleles carry the same instruction, then the individual will have that characteristic. For example, if both alleles are for 'blue eyes', the individual will have blue eyes.

If the alleles are different, then one has to 'win'. The allele that wins is the **dominant** allele. Its characteristic is the one that shows in the individual.

The allele that 'loses' is still present in the individual's chromosomes, but the characteristic that it carries does not show. It is the **recessive** allele.

The outcome of a contest between different alleles can always be predicted – the same one always dominates.

The allele for brown eyes is always dominant over the allele for blue eyes. This is one reason why there are more brown-eyed people than blue-eyed people.

**Did you know?**

Red–green colour blindness is common in males. It is inherited on the X chromosome from the mother. Although this is a recessive gene, the Y chromosome does not have an equivalent gene to dominate it.

## Genetic diagrams

The results of inheriting dominant and recessive alleles can be shown on a diagram called a Punnett square. There are certain rules for drawing the Punnett squares:

- Dominant alleles are shown by uppercase (capital) letters.
- Recessive alleles are shown by lowercase (small) letters.
- The same letter is used for the dominant and recessive alleles. For example, the allele for tongue-rolling is dominant and is shown as T. The allele for non-tongue-rolling is recessive and is shown as t.

A Punnett square shows all the possible combinations of alleles that can be inherited from the parents.

The Punnett square opposite shows how eye colour is inherited. The letter **B** represents the allele for brown eyes and the letter **b** the allele for blue eyes.

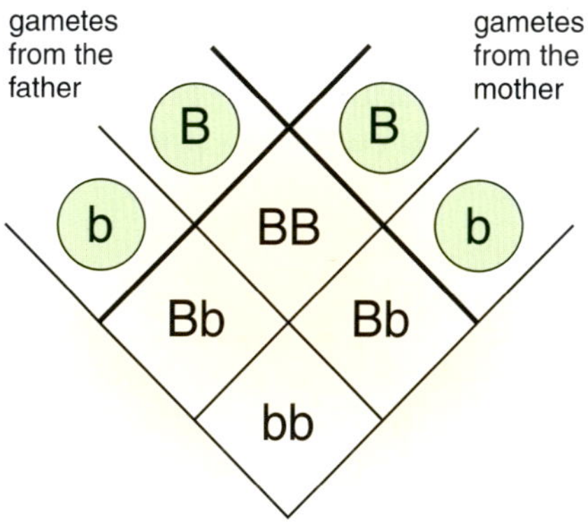


In this example:

- Both parents have brown eyes.
- Both parents carry the dominant allele for brown eyes.
- Both parents also carry the recessive allele for blue eyes.

The diagram shows the four possible combinations of alleles that the offspring could have. This does not mean that if there were four children there would be one with each combination of alleles. The Punnett square shows the chance or probability of a particular combination of alleles occurring.

**b In the example shown, what are the chances that an offspring will have the alleles Bb?**

Sometimes diseases can be passed from parent to offspring by faulty alleles. Diseases passed on in this way are called **inherited disorders**. Cystic fibrosis and Huntington's disease are both examples of inherited disorders.

## Questions

**1** Copy and complete the sentences. Use words from the list.

**allele** **dominant** **gene** **recessive**

The part of a chromosome that controls a characteristic is a ____________.

Different forms of the same gene are called ____________s.

If an individual has two different alleles for a particular characteristic, the ____________ one controls the characteristic.

**2**

The diagram shows how the colour of a rabbit's fur depends on the alleles that the rabbit inherits. The allele for black fur, B, dominates the allele for white fur, b.

The male black rabbit has the genes BB for fur colour.

Explain why all of its offspring have black fur.

### Key ideas

- *Different forms of the same gene are called alleles.*
- *When an individual has two alleles for the same gene, the characteristic of the dominant allele shows.*

# 2:8 Changing the message

## Mutation

Your cells contain several genes that control the production of a substance called melanin. Melanin gives your skin its colour. (It is this pigment that gives you a suntan.)

The photograph opposite shows a blackbird that cannot make melanin. Animals that cannot make melanin and have no pigment in their skin are described as albino.

Albinos cannot make melanin because one (or more) of the genes that control its production has changed. A change in the chemical structure of a gene is called a **mutation**.

Not all blackbirds are black! This is an albino blackbird.

## Increasing mutations

Many mutations occur by chance. The chance of mutations occurring can be increased by exposure to factors such as ionising radiation and some chemicals.

- Ionising radiation includes ultraviolet light, X-rays, and radiation from radioactive substances.
- Chemicals that can increase the chance of mutations occurring include benzene, and the hydrocarbons in tobacco smoke.

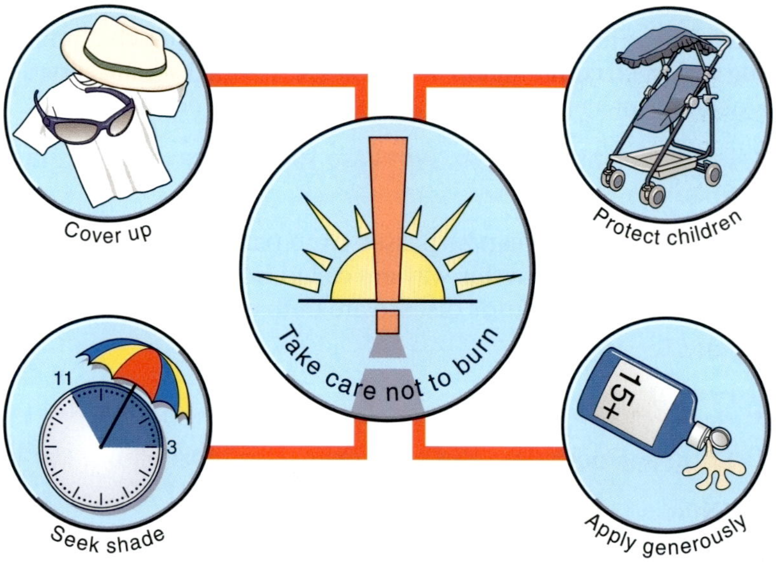


It is important to protect your skin from being damaged by the Sun.

**Harmful mutations**

Some forms of cancer are caused by gene mutations. For example, mutations in skin cells can cause the cells to grow and divide in an uncontrolled way. This is **skin cancer**.

Being exposed to the Sun's harmful ultraviolet radiation can cause skin cancer. People who spend hours trying to get a suntan are increasing their chances of getting skin cancer. This is why doctors advise people to follow the Sun safety code, shown here.

**a Why do black- or brown-skinned people rarely get skin cancer?**

## Talent or training?

Not all characteristics are passed on from parents. You cannot just be born with all that it takes to become a world-class musician – it depends on years of practice as well as having natural talent. This is because the **environment** in which you live and grow affects some characteristics.

The characteristics of individuals may be the result of:

- inherited causes
- environmental causes
- a mixture of both.

**b Which of the following characteristics are inherited, which are environmental and which are a mixture?**

**i height;**
**ii pierced ears;**
**iii eye colour.**

**Modifying a child's inheritance**
Scientific studies show that there is a link between intelligence and birth weight. Bigger babies make brainier people! A number of environmental factors are known to affect birth weight.

- If the mother smokes during pregnancy, birth weight is reduced by up to 200 g.
- If a mother is subjected to passive smoking during pregnancy, birth weight is reduced by up to 100 g.
- Poor diet during pregnancy can also reduce birth weight.

It is therefore important that a pregnant woman avoids breathing in cigarette smoke and has a balanced diet. If she doesn't, she could be harming her unborn child for the whole of its life.

**c How can a mother avoid damage to her fetus by passive smoking?**

It takes more than natural talent to become a world-class gymnast.

**Did you know?**
Scientific studies have shown that smoking while pregnant can cause spontaneous abortions, stillbirth and sudden infant death.

## Questions

**1 a What is a mutation?**

**b Name three causes of mutations.**

**2 Explain why doctors advise parents not to expose babies to direct sunlight.**

**3 The table shows how birth weight is linked to smoking.**

| Number of cigarettes smoked per day | Average birth weight in kg |
|---|---|
| 0 | 3.00 |
| 10 | 2.99 |
| 20 | 2.97 |
| 30 | 2.91 |
| 40 | 2.80 |

**a Plot the data from the table on graph paper. Draw a line of average birth weight against number of cigarettes smoked per day. Draw the best line.**

**b Describe the trend that the graph shows.**

**c A woman smokes 25 cigarettes a day while she is pregnant. How much smaller will her baby be compared with its likely weight if she didn't smoke?**

## Key ideas

- *A gene mutation is a change in the chemical structure of a gene.*
- *Mutations occur by chance.*
- *Exposure to ionising radiation and some chemicals increases the chances of mutations occurring.*
- *Environmental factors can modify some inherited features.*

# 2:9 Changing genes

In recent years scientists have found ways of cutting genes from bacteria and inserting them into plants grown as crops by farmers. This can increase yields, and produce plants that are resistant to pests and diseases so that farmers need to use fewer chemicals.

Crops which have had their genes changed in this way are called **genetically modified** crops (**GM** crops).

Some people are against the use of GM crops. They are concerned about possible harm to the environment and the unknown long-term effects of adding and replacing genes. Other people think that the benefits of having crops that can survive in drought and flood conditions, as well as giving improved nutrition, outweigh the possible disadvantages.

**a Give two advantages that GM crops can bring.**

## More beans

Soya beans have been genetically modified so that they are resistant to a particular weedkiller. This allows farmers to spray the crop with the weedkiller. Weeds are destroyed but the crop is not affected.

The advantages to farmers are:

- it is easier and cheaper to control weeds
- yields are higher because there are fewer weeds taking nutrients from the soil
- they do not need to till the ground before planting the crop.

**b How could not tilling the ground be an environmental benefit?**

Genetically modified soya plants produce more soya beans than do normal plants.

## Amazing maize

The corn borer insect causes a lot of damage to maize crops, reducing the yield. Maize has been genetically modified so that it produces a protein that kills this insect. Killing the insect prevents it from damaging the corn and stops it from reproducing. Farmers who grow genetically modified maize do not need to use chemicals to control the pest.

**c Explain why using GM corn can help to reduce chemical pollution in the environment.**

The corn borer insect can devastate maize crops.

## Cool cotton

Like maize, cotton crops can be damaged by insects. The insect larvae feed on the plants. Genetically modified cotton produces a protein that reduces the survival rates of the larvae. The larvae do less damage to the crop and so less insecticide is needed.

**d Cotton plants can be genetically modified to produce crops of different colours. What is the advantage of this?**

The cotton bollworm can destroy cotton crops.

## For and against

GM crops have several potential benefits. In particular they could help to improve the nutrition of people in the world's poorest countries. However, there are some environmental concerns. Here are just some of these concerns.

- Plants that are resistant to weedkillers could become difficult to control.
- Genes from genetically modified crops may be transferred to similar crops or to wild plants.
- The chemicals that plants produce to kill pests could kill harmless insects and pass through the food chain to kill birds.

The only GM crops that are allowed to be grown in the UK are those used for trials to find out about their effects on the environment. All foods that contain GM ingredients have to be labelled.

**Did you know?**

Most crops are grown from F1 hybrid seeds. These are seeds produced by crossing particular male and female plants. The characteristics of the offspring are known to be reliable and desirable.

## Questions

1 Copy and complete the sentences. Use words from the list.

bacteria    genes    genetically modified

GM crops have been ______________. This is done by adding ______________ that have been taken from ______________.

2 Describe the advantages of growing GM soya beans.

3 What are the possible harmful effects of growing GM crops?

## Key ideas

- *GM crops can give high yields, survive in adverse weather conditions and resist attack by pests and disease.*
- *The long-term environmental effects of growing GM crops are unknown.*

# 2:10 Survival

The photograph shows part of the Sahara desert. You cannot see any animals or plants in the photograph. Desert animals normally hide during the day. These animals have many special **features** that enable them to survive in extreme conditions.

The Sahara desert is one of the hottest environments on Earth.

## Big ears

Most mammals have body temperatures similar to humans – about 37 °C. Temperatures in deserts are often in the range 40–60 °C.

Instead of trying to conserve heat, like arctic animals, many desert animals have **adaptations** for keeping cool.

The fennec, the jackrabbit and the bandicoot all have big ears. These ears act to get rid of excess heat from the blood. They receive warm blood from the body, radiate excess energy to the surroundings, and then return cooler blood to the body.

These animals can increase the rate of heat loss by increasing the blood flow through the ears.

The bandicoot has large ears to help it lose heat.

## Cacti

Cacti range in size from the small ones that people grow in pots on windowsills to huge ones such as the one shown in the photograph below. Cacti are adapted to growing in dry places. Those in the photograph seem to be growing on bare rock.

Look at the photographs of the cacti. You will notice that cacti have swollen stems and that they appear to have no leaves. Cactus stems are swollen to help them store water in order to survive dry conditions.

It is not quite true to say that cacti have no leaves. The leaves have not been lost; they have evolved into small spines. Because cacti have very small leaves (spines) they lose very little water through **evaporation**.

**Did you know?**
Few large animals can survive desert conditions. The camel is an exception; fat stored in its hump releases water in respiration.

Some cacti grow to enormous sizes.

Some cacti appear to grow out of bare rock.

**a How else does having spines instead of leaves help cacti to survive?**

Plants manufacture food by photosynthesis. Light is absorbed by chlorophyll and is used to convert carbon dioxide and water into sugars. Because cacti have such small leaves, they need to use their stems to produce food.

**b Look at the photographs of cacti. How can you tell that photosynthesis occurs in the stems of these plants?**

Cacti have other adaptations that help them to survive in hot dry conditions. Some cacti have very long roots that reach down to any water stored far beneath the surface. Other cacti have very wide-spreading shallow root systems to take up as much water as possible when it rains. Cacti stems and spines have a waxy layer to reduce water loss.

## Living under water

Fish are different to animals that live on land in a number of ways.

- Gas exchange is carried out in **gills** instead of lungs. Gills are designed to absorb oxygen that is dissolved in water and to release carbon dioxide into water.
- Fish have a **streamlined** shape – the resistive forces are much greater when moving through water than when moving through air.
- The skin is covered in **scales** that overlap, with a layer of mucus on top. This reduces the resistive forces still further.

**c How does gas exchange in gills differ from gas exchange in lungs?**

Fish have many adaptations to life in the water.

## Questions

1 Explain why polar bears have small ears and bandicoots have big ears.

2 Copy and complete the sentences about cacti. Use words from the list.

**adapted roots spines water**

Cacti are ______________ to live in dry places. Their ______________ penetrate deep underground. Their leaves have evolved into ______________ that have a very small surface area and their stems are swollen to store ______________.

3 Marram grass inhabits sand dunes. In wet conditions its leaves are flat, but in dry conditions they roll up. How does this help the marram grass to conserve water?

4 a Describe three ways in which a fish is adapted to live in water.

b Explain how each adaptation helps the fish to survive.

## Key ideas

- *Animals and plants are adapted in different ways to be able to survive in their environment.*
- *Desert animals have large ears to radiate energy to their surroundings.*
- *Cacti have small leaves and either deep roots or wide-spreading shallow roots.*
- *Fish have gills for gas exchange, and scales and a layer of mucus for streamlining.*

# 2:11 Populations

Farmers spend a lot of time and effort getting rid of weeds from their crops. Weeds compete with crop plants for **resources** such as light, water and nutrients.

Weeds are not the only **competition** for crop plants. Plants also compete with each other. Farmers need to know how far apart to plant their crop plants to get the best yield. If they plant them too close, there is too much competition for scarce resources and the yield is reduced.

**Farmers remove weeds so that they do not take valuable nutrients from their crop plants.**

## Sea birds

The photograph shows part of a cliff face. The sea birds that live here compete for space to breed and for food.

If the **population** increases, there is more waste from faeces. This can cause disease that spreads rapidly because of overcrowding.

The population of a species in a habitat is limited by the resources that are available. If there is not enough food or space to support the population then some individuals die, reducing the population to a size that can be supported by the habitat.

**This cliff face is packed with nesting birds. There will be a lot of competition for food.**

**a Why does disease spread more rapidly if there is overcrowding?**

## Predator and prey

Toothed whales are known to catch leopard seals. Animals that kill and eat other animals are called **predators**; the animals they eat are called **prey**. The toothed whale is the predator. The leopard seal is its prey. Predators usually have more than one type of prey.

Look at the diagram of a food web. This shows that Weddell seals and large fish are also the prey of toothed whales.

**Killer whales prey on seals.**

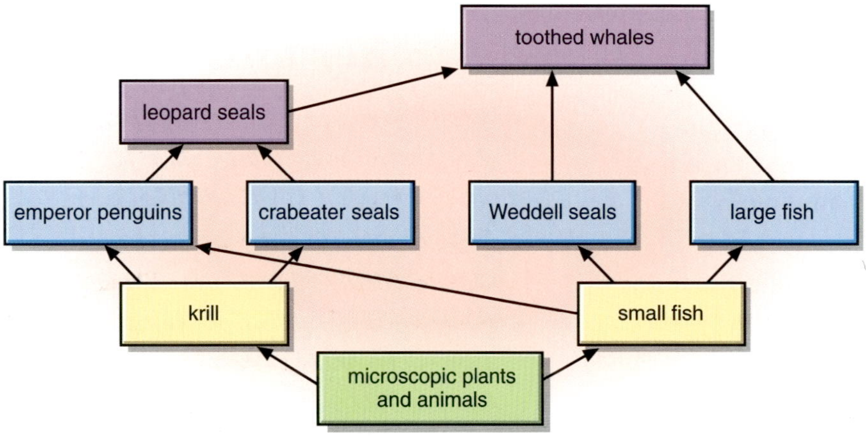


**A food web tells you what eats what. The arrows show the direction of energy flow – from the prey to the predator.**

**b The population of Weddell seals is almost wiped out by disease. Explain how this is likely to affect the population of toothed whales.**

## Population size

The size of the population of predators depends on the food supply.

- If there is more food the number of predator increases.
- More prey are eaten so there is less food and many predators starve.
- Fewer prey are eaten so they reproduce and their numbers increase again.

The populations of many predators and prey increase and decrease in this way.

Owls are predators of voles. The chart shows how the number of owls and voles in a forest varies over a ten-year period.

You can see from the chart that an increase in the number of voles is followed by an increase in the owl population. More voles are eaten, so their population goes down and then the owl population goes down.

**c Explain why there is a large drop in the vole population between years 5 and 6.**

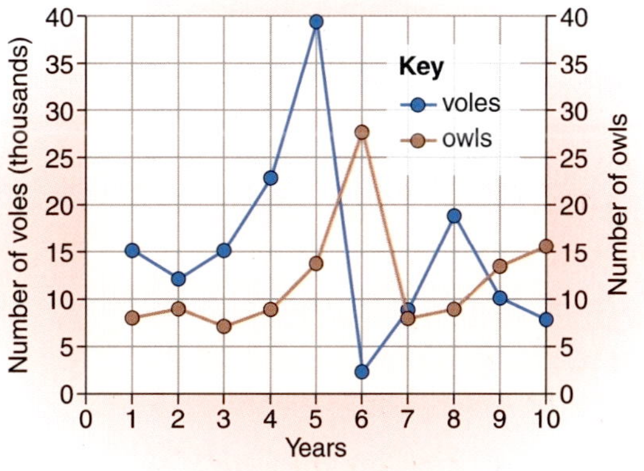


**Owls are predators. Voles are just one of their prey.**

## Questions

**1** **Why do farmers pull out the weeds that grow between potato plants?**

**2** **A sparrowhawk eats blue tits. Which is the predator and which is the prey?**

**3** **The table below shows how the numbers of roach (a type of fish) and water fleas in a lake changes over seven years.**

| Year | Number of roach in lake | Number of water fleas in each litre of water |
|---|---|---|
| 1 | 70 | 120 |
| 2 | 250 | 5 |
| 3 | 20 | 80 |
| 4 | 100 | 40 |
| 5 | 30 | 260 |
| 6 | 190 | 20 |
| 7 | 75 | 40 |

**a** **Draw a graph to show how the populations change.**

**b** **Describe the pattern in the change in the population of water fleas.**

**c** **Explain how the graph shows that the roach are predators of water fleas.**

**d** **When the number of water fleas is very low the roach still survive. Suggest how they do this.**

### Key ideas

- *The size of a population is limited by the resources that individuals compete for.*
- *Animals that eat other animals are called predators. The animals that they eat are called prey.*
- *If the population of prey increases, there is more food for the predator. The predator population increases. In turn, this causes the prey population to decrease.*

# 2:12 Evolution

Life is full of variation. New individuals with a different set of characteristics are being produced all the time. Individuals within a species vary because of the different combinations of alleles that they inherit. Mutations create new characteristics that may make an individual more suited or less suited to its environment.

In the natural world, individuals that are not suited to their environment die and do not reproduce. Life is harsh!

**a A mutation can cause an animal to be born albino. It has no natural colouring so it appears white. Suggest why few albinos survive to adulthood.**

Dinosaurs died out when their environment changed and they were no longer suited to it.

## Survival of the fittest

In the UK population, most babies survive through childhood to become adults. However, this is not the case in some countries and it is certainly not true in the animal kingdom. Young animals are often easy prey for predators. In order to survive they need to be able to:

- sense danger from predators
- react and move quickly to remove themselves from the danger of predation.

Only those animals that can do this will survive, become adults and produce offspring. This is how **natural selection** works: only the fittest survive.

**b Explain why fewer babies die in the UK nowadays compared with a hundred years ago.**

Young animals are particularly vulnerable to predators. Can you see which cub is watching for danger?

## The evidence

Fossils show life as it was millions of years ago. The bones and shells of animals, as well as the outlines of leaves and plants, are preserved in rocks as fossils. By looking at fossils, scientists can tell which species have survived unchanged, which have become **extinct** and which have changed to adapt to a changing environment. Those that have changed are said to have **evolved**. The process is known as **evolution**.

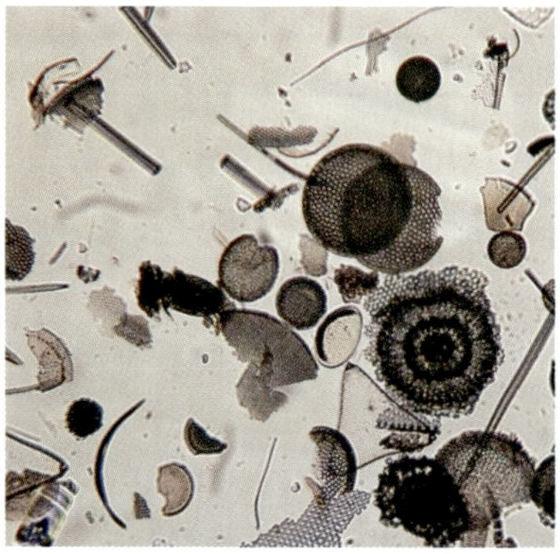

Fossils are relics of animals and plants that lived millions of years ago.

**Did you know?**

Evidence from fossils suggests that the earliest forms of life were bacteria-like organisms that existed 3000 million years ago.

## The peppered moth

Before the Industrial Revolution, the peppered moth was well camouflaged on tree trunks.

The peppered moth lives on the bark of trees. Two hundred years ago peppered moths were pale and so were camouflaged against pale tree trunks. They were adapted very well to their environment.

Because of a mutation, a variant of the peppered moth appeared. The variant was black. These moths were not adapted to their environment because they stood out against the pale bark of the trees. Birds could easily spot them and eat them. Very few of these variants survived to become adults and reproduce.

However, when the UK became industrialised and a lot of coal was burned, tree trunks became blackened with soot. The pale peppered moths were no longer camouflaged but the black ones were. The black moths were now more suited to their environment. They survived and reproduced whereas the pale moths were easily spotted and eaten.

This is how peppered moths evolved from a pale colour to a dark colour.

**c What would have happened to the peppered moth if a black variant had not evolved?**

When soot turned the tree trunks black, the peppered moth was no longer camouflaged.

Dark peppered moths were well camouflaged on soot-covered tree trunks and so survived to reproduce.

## Questions

**1** The following statements summarise how species may change by natural selection. Place the statements in the correct order to show how this process takes place.

- Organisms that survive and breed will pass on their genes to the next generation.
- Individual organisms may have different characteristics.
- Organisms with characteristics most suited to the environment are more likely to survive and breed.
- Organisms may die because of predation, disease and competition.

**2 a** What is a fossil?

**b** Evidence from fossils suggests that giraffes used to have much shorter necks than they do nowadays. Explain how long-necked giraffes could have evolved from short-necked giraffes.

**3 a** Explain how a new characteristic can appear in a species.

**b** What happens to an individual with a new characteristic if it is not suited to its enviuronment?

**c** How does natural selection ensure that only the fittest individuals survive and breed?

### Key ideas

- *New species evolve from variants that are better suited to their environment.*
- *Much of the evidence for evolution comes from fossil records.*
- *Natural selection ensures that only those individuals that are most suited to their environment survive and breed.*

# 2:13 Selecting the best

## Tasty foods

The shelves of our supermarkets are stacked with a wide choice of high-quality foods. This food has been produced from plants and animals bred by farmers.

Farmers choose plants and animals that have useful characteristics and then use them for breeding. This means that the useful characteristics are passed on to new individuals. Choosing the plants and animals used for reproduction is called **selective breeding**. In this way new varieties of plants and animals are being produced.

**Nowadays we can buy a huge variety of high-quality produce from our supermarkets.**

## Growing more wheat

Wheat is a very important crop. It is ground into flour to make bread, cakes and other foods. Modern varieties of wheat have characteristics that are more useful to farmers than those of older varieties.

New varieties of wheat have been produced by crossing high-yielding wheat with dwarf wheat and then using only the seed from those plants that have the desired characteristics to grow more seed. This is because:

- they have short stems so less energy is wasted in growing straw
- they use a greater proportion of their energy to produce seeds.

**Modern varieties of wheat have many advantages over old varieties.**

**a Why do farmers prefer to grow wheat that has short stems rather than wheat that has long stems?**

## More milk

The black-and-white cows that you often see in fields are a breed called Friesian. This breed is used by farmers to produce milk. Selective breeding over generations has produced Friesian cows that give high yields of milk.

Few cows have any physical contact with a bull to reproduce. Instead, they are fertilised by **artificial insemination**.

1. Sperm are collected from the testes of a bull.
2. The sperm are squirted into the vagina of a cow using a syringe.
3. In this way the sperm from one bull can be used to fertilise the ova of thousands of cows.

**Selective breeding has produced Friesian cows that produce a high milk yield.**

The sperm from bulls that produce offspring with high milk yields is used to fertilise Friesian cows. The cows are also chosen for their high milk yields. This process of selective breeding gradually improves the milk yield in the population of Friesian cows. Selective breeding can also be used to eliminate undesirable characteristics.

**b What are the advantages of using artificial insemination to fertilise the ova of cows?**

**Did you know?**
Paintings on the walls of Egyptian tombs show that selective breeding of dogs has been carried out for at least 4000 years.

## One sheep or two?

Sheep are farmed for their meat. Lambs have a higher market value per kilogram than ewes. Their meat is more tender and tasty. Some sheep have an allele that makes them more likely to have multiple births, producing twin lambs rather than a single lamb.

Farmers choose to breed lambs from these sheep. By continuing to breed from sheep that have the multiple-birth allele, the allele for single birth becomes less common and is gradually eliminated.

**c When sold as meat, why do lambs have a higher market value than ewes?**

This ewe produces more meat than one that only has one lamb at a time.

## Questions

1 Copy and complete the sentences. Use words from the list.

**artificial    breed    characteristics    selective**

Friesians are an example of a ______________ of cow. They were produced by ______________ breeding. This means that only animals with chosen ______________ are allowed to breed.

2 List three characteristics of modern wheat that make this variety more useful than wild wheat.

3 Dairy Shorthorns are a breed of cow that were once used for milk production. The graph below shows the average amount of milk produced per cow in a year for Dairy Shorthorns and Friesians.

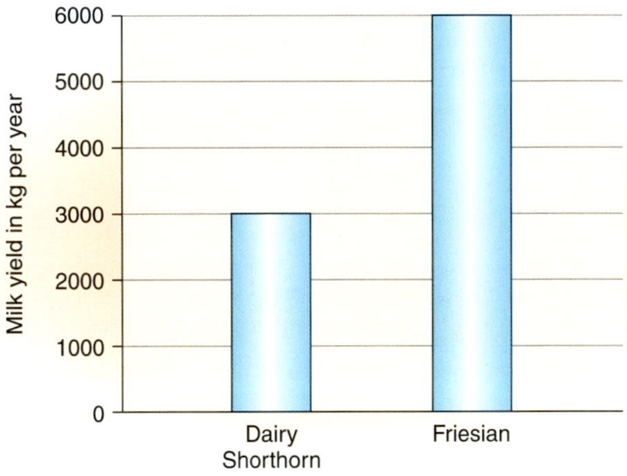


Milk yields from Friesian and Dairy Shorthorn cows.

a How much milk is produced by:

i Dairy Shorthorns;
ii Friesians?

b Explain why farmers changed to using Friesians.

## Key ideas

- *Selective breeding is used to produce new varieties of plants and new breeds of animals.*
- *Using modern varieties of plants and breeds of animals has increased food production.*
- *Choosing the animals and plants that are used for breeding is known as artificial selection.*

# 2:14 Pollution

By living on this planet, we pollute it. As the human population grows, it produces more **pollution**. The more developed and industrialised countries are responsible for most of the world's pollution. These countries burn **fossil fuels** in vast quantities, which is changing the composition of the atmosphere.

Coal-burning power stations produce a lot of pollution.

## Acid rain

The photograph shows a coal-burning power station. Coal is burned to release energy. The main element in coal is carbon. During burning the coal is oxidised to form carbon dioxide. Carbon dioxide dissolves in rainwater and makes it slightly acidic.

Most kinds of coal also contain sulphur. When sulphur burns it is oxidised and forms sulphur dioxide. Sulphur dioxide dissolves in rainwater and makes it more acidic. Breathing in sulphur dioxide can damage your lungs.

**Acid rain** has a pH of 4 or less. It has very damaging effects on the environment.

**a How is acid rain damaging some stone buildings?**

### Cars

Cars also burn fossil fuels. Petrol and diesel are hydrocarbons. The exhaust gases from a car include:

- carbon dioxide and water, produced when hydrocarbons burn
- carbon monoxide, a poisonous gas produced during incomplete combustion
- sulphur dioxide – this can be reduced by using low-sulphur petrol and diesel
- nitrogen oxides, which also dissolve in rainwater and make it more acidic.

Many cars are now fitted with catalytic converters, which reduce the amount of nitrogen oxides in the exhaust fumes.

**b What are the main elements in hydrocarbons?**

Acid rain damages the young leaves at the crown of the tree.

### Damage to trees

The photograph shows trees that have been damaged by acid rain. These trees have a condition called 'crown loss'. Their young leaves, those near the top of the tree, are killed by the acid rain. Acid rain also affects the covering of the older leaves. This makes it easier for disease organisms to attack the tree.

The young root hairs of the trees are damaged when acid rainwater gets into the soil.

**c What is the job of root hairs? How will the tree be affected if its root hairs are damaged?**

**Did you know?**
Half of the trees in the Black Forest in Germany have been damaged by acid rain.

**How acid rain affects freshwater organisms**

When acid rain runs into streams and lakes it makes the water more acidic. As the water gets more acidic, fewer animals and plants can survive. The chart shows the level of acidity that some animals can tolerate.

**d i Which fish tolerates acid water the best?**

**ii What is the lowest pH of water that snails can live in?**

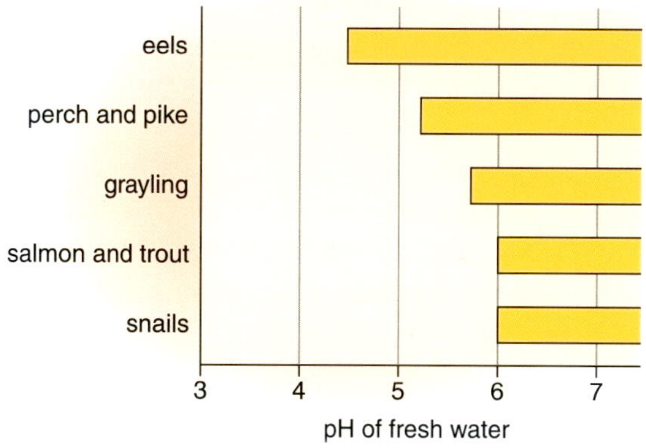


**Different organisms can tolerate different ranges of pH. All the animals listed except snails are fish.**

## Reducing air pollution

Most air pollution is caused by burning fossil fuels. It can be reduced by:

- using cars less so that less fuel is burned
- burning fuel more efficiently
- burning fuels that cause less pollution.

This involves taking more journeys on foot or bicycle, and using public transport when possible. Drivers of small fuel-efficient cars now pay less tax than those who drive cars that use more fuel and cause more pollution.

## Questions

**1 Which gases in motor vehicle exhaust fumes contribute to acid rain?**

**2 Which gases in motor vehicle exhaust fumes are removed by a catalytic converter?**

**3 The map shows the percentage of trees damaged by acid rain in European countries.**

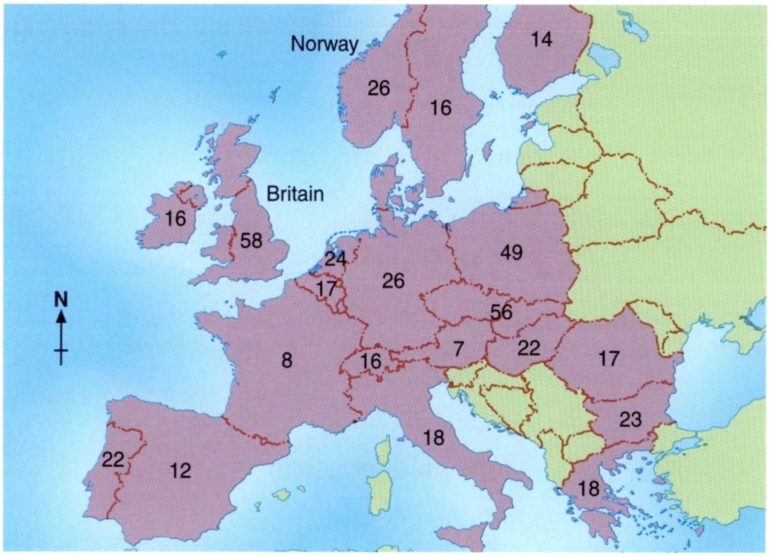


**The numbers show the percentage of trees damaged by acid rain in each country.**

**a Work out the average percentage of trees damaged in European countries.**

**b If you looked at 1000 trees in Norway, how many would you expect to be damaged by acid rain?**

**c Ireland has a lower percentage of trees damaged by acid rain than England. Suggest an explanation for this.**

## Key ideas

- *Burning fossil fuels such as coal and oil releases sulphur dioxide into the atmosphere.*
- *Nitrogen oxides are also released when petrol is burned in car engines.*
- *Sulphur dioxide and nitrogen oxides dissolve in rainwater and make it more acidic.*
- *Acid rain kills the young leaves of trees and may kill the root hairs. If the water in lakes and rivers becomes too acidic, plants and animals cannot survive.*

# End of module questions

## Low demand

**1** A normal human cell contains 46 chromosomes.

How many pairs of chromosomes are there in a normal human cell?

**A** 23

**B** 46

**C** 64

**D** 92

**2** Sexual reproduction always involves:

**A** ejaculation

**B** fertilisation

**C** menstruation

**D** pollination

**3** A female sex cell is known as:

**A** cytoplasm

**B** an egg

**C** a sperm

**D** a tail

**4** The menstrual cycle describes changes in the:

**A** penis

**B** testes

**C** uterus

**D** vagina

**5** In which part of a cell are the chromosomes?

**A** the cell membrane

**B** the cell wall

**C** the cytoplasm

**D** the nucleus

**6** A mutation can change a:

**A** finger

**B** gene

**C** leg

**D** cell

**7** When a plant is genetically modified, changes take place to its:

**A** cytoplasm

**B** cell membrane

**C** cell wall

**D** genes

The graph shows how the population of fish in a pond changes.

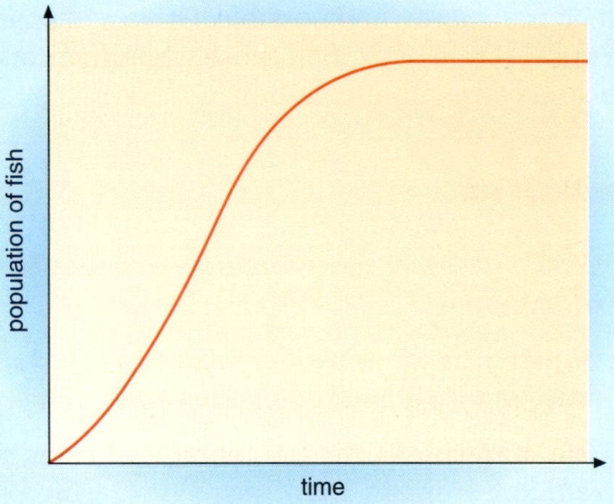


**8** Look at the graph. It shows that the population of fish:

**A** rises and then falls

**B** rises to a steady value

**C** falls then rises

**D** falls to a steady value

**9** The population of fish could be controlled by:

**A** competition for food

**B** competition for air

**C** competition for speed

**D** predators

**10** Which activity can release acidic gases into the atmosphere?

**A** fish farming

**B** photosynthesis

**C** selective breeding

**D** burning fossil fuels

11 Changes in the body at adolescence are caused by:

**A** gametes

**B** hormones

**C** mutations

**D** radiation

12 The sex of a human is determined at:

**A** adolescence

**B** birth

**C** fertilisation

**D** ovulation

## Standard demand

13 The haploid number is the number of chromosomes in a:

**A** gene

**B** gamete

**C** red blood cell

**D** white blood cell

14 Which graph shows how world pollution changes with increasing human population?

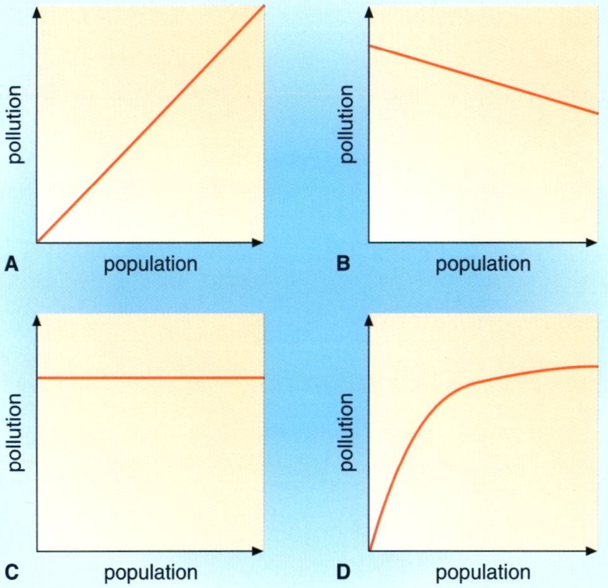


15 The growth of facial hair on a human male is controlled by:

**A** oestrogen

**B** penisterone

**C** progesterone

**D** testosterone

16 An individual is male if its sex chromosomes are:

**A** XX

**B** XY

**C** YY

**D** YZ

17 An individual inherits characteristics from its father through the:

**A** egg

**B** red blood cells

**C** sperm

**D** white blood cells

18 Asexual reproduction:

**A** begins with fertilisation

**B** involves only one parent

**C** requires two parents

**D** produces offspring with variation

19 Alleles are different forms of:

**A** chromosomes

**B** gametes

**C** genes

**D** hormones

20 The graph shows how the populations of a predator and its prey change.

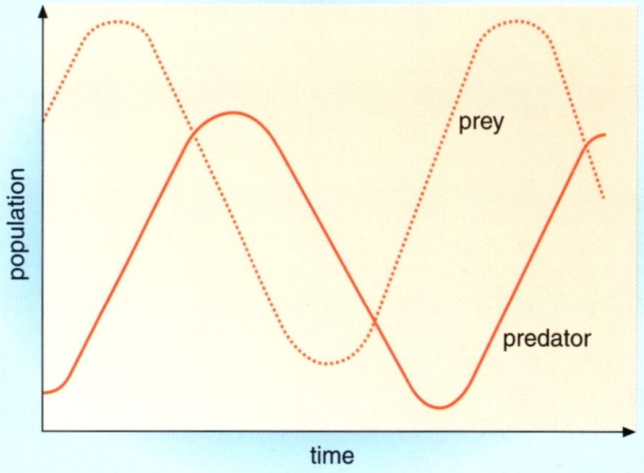


The population of predators rises because:

**A** the population of prey has risen

**B** the population of prey has fallen

**C** the population of prey has remained constant

**D** another predator has eaten the prey

**21** A farmer breeds from his cows that have the highest milk yield.

This is known as:

**A** artificial insemination

**B** cloning

**C** genetic modification

**D** selective breeding

**22** A seal has a thick layer of fat to:

**A** help it to survive in cold water

**B** help it to survive in warm water

**C** deter predators

**D** attract prey

**23** Evidence for evolution comes from:

**A** bones

**B** fossils

**C** relics

**D** teeth

**24** A reason for genetically modifying a crop is so that it:

**A** destroys itself

**B** needs more fertiliser

**C** needs more insecticide

**D** produces a chemical that poisons pests

# Module 3 – Chemical patterns

This module is divided into four parts.

1 All materials are made of atoms. Atoms themselves are made of protons, neutrons and electrons. How are these particles arranged in an atom?

2 The periodic table is a chart showing the names and symbols of elements.

3 The halogen elements make up one family of elements in the periodic table. Within this family there are trends and patterns both in physical and chemical properties.

4 Chemical reactions can take place at widely different rates. What are the factors that affect the rate of a reaction? The patterns in reaction rates can be explained using a collision theory.

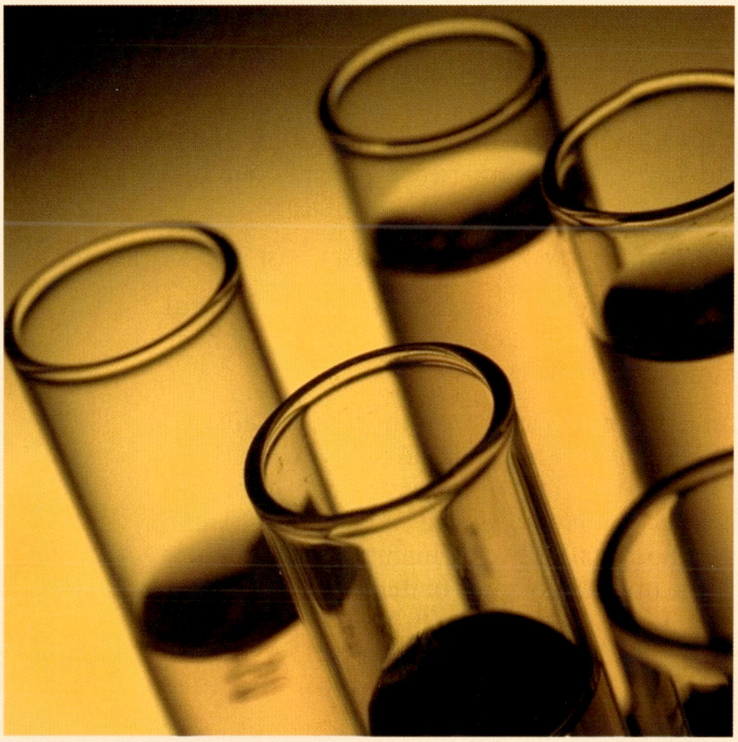

Try these first.

**1** Which of the following are not elements?

air    brass    carbon    copper    oxygen

**2** Which of the following is a compound?

chlorine    salt    sodium

**3** Match up the names of elements with their symbols. Choose symbols from this list.

C    Cu    K    Na    P    S

a sulphur _______ ;  b phosphorus _______ ;  c potassium _______;

d sodium _______ ;  e carbon _______ ;  f copper _______.

**4** What are the chemical symbols for hydrogen, calcium, magnesium and iron?

The diagram shows iron reacting with dilute hydrochloric acid. It makes hydrogen bubbles.

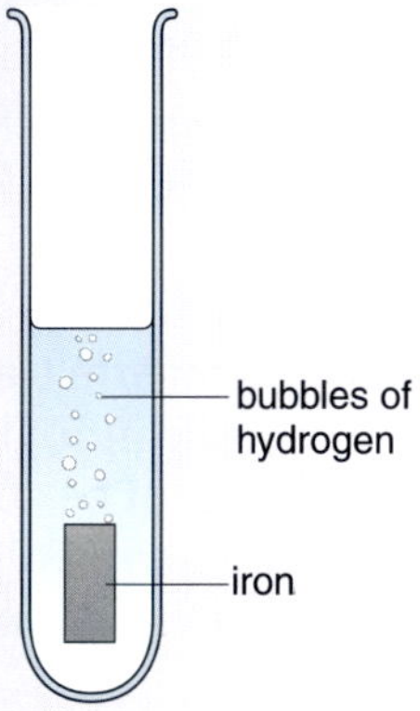


**5** How could you speed up this reaction without changing the chemicals?

**6** How would you be able to tell when the reaction was faster?

**7** What is formed when elements combine?

**8** Water is made up from hydrogen and oxygen. What does the formula $H_2O$ tell us?

# 3:1 Atoms

## Smaller and smaller

If you start with a block of iron, you could cut it into two smaller pieces. If one of the pieces is then cut again into two, two smaller pieces of iron are produced. If you were to keep on cutting the pieces into two, eventually you will finish up with the smallest piece of iron that can exist.

The smallest piece of iron that can exist is called an iron **atom**.

If you were to do the same thing with a piece of copper, you would eventually finish up with an atom of copper.

An atom of iron and an atom of copper are different in mass and size. A piece of iron is made up only of iron atoms and a piece of copper is made up only of copper atoms. There are about 100 known elements. Each element is made up only of atoms of that element.

**a** What is a piece of sulphur made up from?

**b** Give two differences between an iron atom and a sulphur atom.

**How big is an atom?**
Atoms are very small indeed. Atoms cannot be seen using ordinary microscopes. The smallest atom is a hydrogen atom.

## Where did the idea of atoms come from?

**Democritus** lived in Ancient Greece about 2400 years ago. He thought that all materials were made up of tiny particles. He called these particles atoms.

His ideas were not based on any experimental results. He believed that different materials would be made of different atoms. These atoms were themselves indivisible.

**Aristotle**, a most powerful man in Ancient Greece, had different views to Democritus and did not believe in atoms.

**c** Why did people stop believing in atoms?

Democritus' ideas were largely forgotten but not completely.

**Did you know?**
The word atom comes from the Greek word for indivisible.

## The idea of atoms continued

About 500 years ago Pierre Gassendi revived the idea of atoms. He stated that atoms cannot be created nor destroyed, that they have weight and cannot be subdivided. He also believed that atoms could be joined together to form groups of atoms. He called these **molecules**. Again he provided no evidence to support his ideas.

## John Dalton

In 1808 Dalton introduced his atomic theory. Unlike the people mentioned above, he had some results from experiments to support his theory.

The main points in his atomic theory were:

- All matter consists of tiny particles called atoms.
- All atoms are indestructible and unchangeable.
- All atoms of the same element have identical weights.
- When atoms react, their atoms combine in simple whole-number ratios.

**d** Why were other scientists more likely to support Dalton's ideas on atoms than those of Democritus?

John Dalton.

## Questions

**1** Copy and complete the following sentences. Use words from the list.

atoms destroyed ratios same split

In his atomic theory of 1808, Dalton said that all matter is made up of tiny particles called _______. These particles cannot be _______ up or _______. All of the atoms in an element are the _______. Atoms combine together in simple whole-number _______.

**2** Dalton produced a chart of elements (right). He represented each element with a symbol. Use this chart to answer the following questions.

**a** The modern name for azote is nitrogen. Write down the names of two gaseous elements in Dalton's table of elements.

**b** Write down the name of a liquid element in Dalton's table.

**c** Which compounds are shown by these diagrams?

i ii iii

**Did you know?**
The last important scientist to be converted to the idea of atoms was Freidrich Ostwald. In 1906, he grudging accepted the existence of atoms.

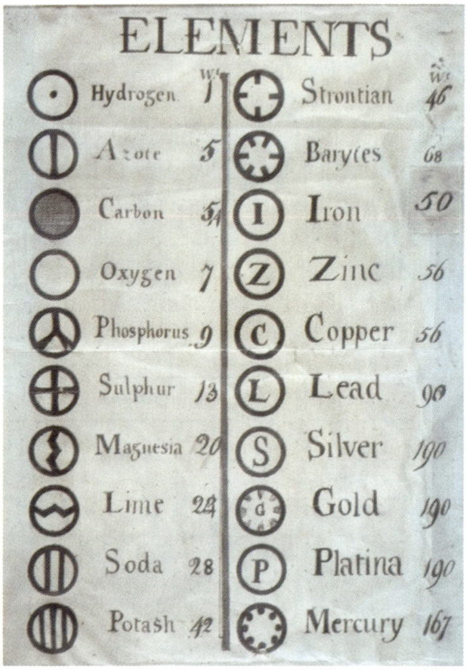


### Key ideas

- *All matter is made up of atoms.*
- *An element contains only atoms of that element.*

# 3:2 Inside the atom

John Dalton believed that atoms could not be split up. In his view, they were like snooker balls.

We now know that atoms can be split up. They are made up of even smaller particles.

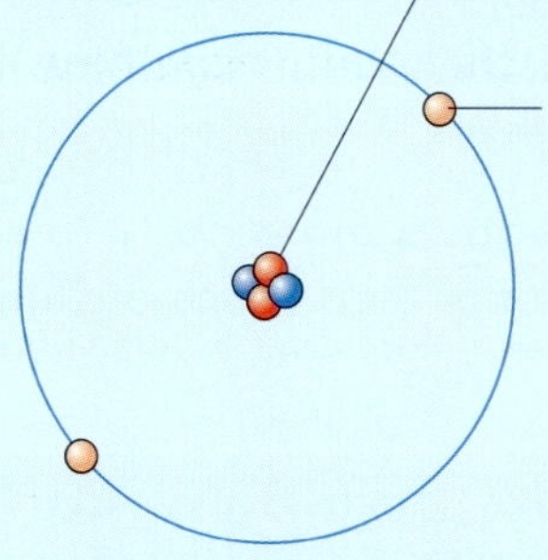


**An atom of helium has two negatively charged electrons whizzing around a nucleus that is made of two positively charged protons and two neutrons.**

## Particles that make up atoms

Atoms are made from just three kinds of particles arranged in a particular way. These sub-atomic particles are called **protons**, **neutrons** and **electrons**.

The same three kinds of particles are used to make all atoms so you could not say, for example, 'this is a copper proton'.

At the very centre of an atom is the **nucleus**. This contains the protons, which carry a positive electrical charge, and the neutrons, which are electrically neutral.

**a Is the nucleus of an atom positively charged, negatively charged or neutral?**

The nucleus is very small compared to an atom.

As well as having different electrical charges, the different particles have different masses. These are too small to be usefully measured in grams, so they are simply compared to the mass of a hydrogen atom. On this model, protons and neutrons both have a mass of 1, but electrons are so small that their mass is usually ignored!

| | Mass | Charge |
|---|---|---|
| **Proton** | 1 | positive |
| **Neutron** | 1 | neutral |
| **Electron** | 0 | negative |

**b Which of the two particles in the table have the same mass?**

**c Which particle in the table has no charge?**

**d Give two differences between a proton and an electron.**

**e What is the mass of the helium atom shown in the diagram?**

**Did you know?**

If you made a model of an atom the size of a football stadium, the nucleus would be the size of a pea on the centre spot.

Atoms of the same element always have the same number of protons in the nucleus. If the number of protons is different, it is a different element!

The number of protons in an atom is called the **atomic number** ($Z$).

Carbon has an atomic number of 6. Any carbon atom must have 6 protons in the nucleus. Atoms are all electrically neutral, so

number of protons = number of electrons.

The positives exactly cancel out the negatives.

## Questions

1 Copy and complete the following sentences. Use words from the list.

atomic electrons nucleus

An atom of lithium (________number = 3) contains three protons and four neutrons in the ________. Moving around this are three ________.

2 The table shows the number of protons, neutrons and electrons in four different atoms. The letters A–D are not chemical symbols.

| Atom | Number of protons | Number of electrons | Number of neutrons |
|---|---|---|---|
| A | 6 | 6 | 6 |
| B | 7 | 7 | 7 |
| C | 6 | 6 | 7 |
| D | 5 | 5 | 4 |

a Which atom has the most particles in the nucleus?

b Two atoms are the same element. Which two?

c Which atom is the heaviest?

d What is the mass of atom D?

3 Magnesium has an atomic number of 12. How many protons and how many electrons are there in a magnesium atom?

4 Scandium is an element with an atomic number of 21. A scandium atom contains 22 neutrons.

a How many protons are there in a scandium nucleus?

b How many particles are there in the nucleus of a scandium atom?

c How many electrons are there in a scandium atom?

d What is the relative mass of a scandium atom?

## Key ideas

- *Atoms are made up from protons, neutrons and electrons.*
- *Protons and neutrons are tightly packed together in the nucleus.*
- *Electrons move around the nucleus.*
- *The number of protons in an atom is the same as the number of electrons.*
- *Two atoms with the same number of protons in the nucleus are atoms of the same element.*

# 3:3 Electrons in an atom

## Electron shells (energy levels)

The electrons whiz round the nucleus of an atom in certain **electron shells**.

The number of electrons is equal to the number of protons. This ensures that the atom is neutral.

**a A hydrogen atom contains one proton. How many electrons does it contain?**

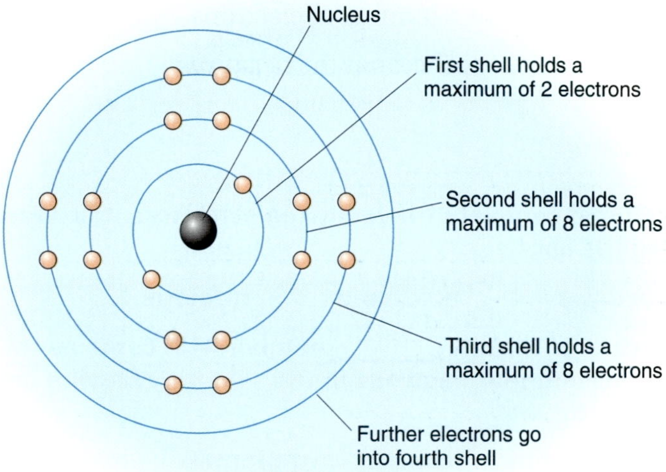


For the first twenty elements, the pattern of filling electron shells is a simple one.

The first shell can only take two electrons, but the next two shells can take eight each. If you know the number of electrons in the atoms, you can work out how they are arranged in the shells.

For these elements the first shell needs to be full before the electrons may enter the second shell.

This arrangement of electrons is called the **electronic structure** of the atom.

## Filling up electron shells

Simple diagrams show the arrangement of electrons in the first twenty elements.

The single electron in a hydrogen atom fits in the first shell.

The two electrons in a helium atom fill up the first shell completely.

A lithium atom has three electrons. Two fill the first shell and the third goes into the second shell.

The pattern continues from beryllium to neon with one electron extra going into the second shell each time.

In the neon atom both the first and second shells are full.

**b A sodium atom contains 11 electrons. How many electrons are there in the first shell, the second shell and the third shell?**

From magnesium to argon, the third shell continues to fill.

**c Neon and argon have similar properties. What is similar about the electronic structures of these two atoms?**

A potassium atom has one more electron that an argon atom. This goes into the fourth shell.

**d A calcium atom has one more electron than a potassium atom. Into which shell does this electron go?**

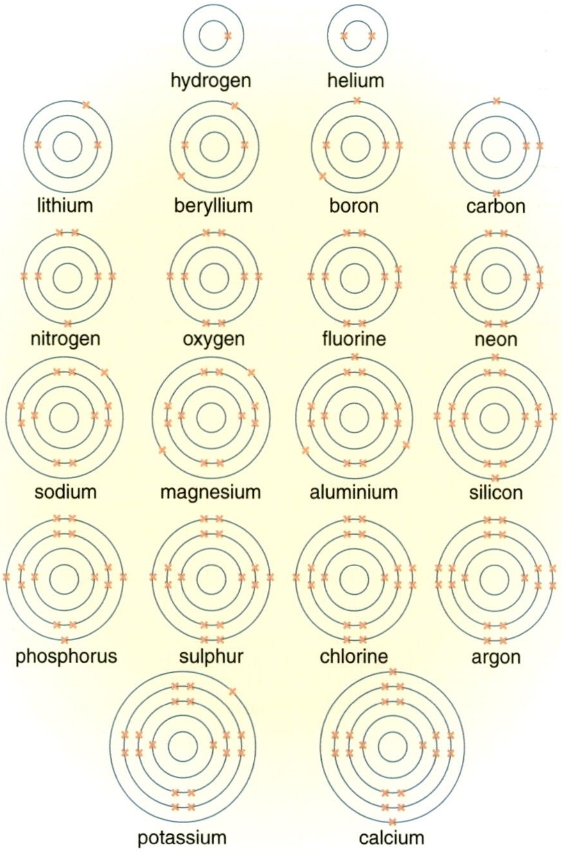

## Questions

1 Copy and complete the sentences using words from the list.

**closest** **eight** **shells** **two**

Electrons occupy ________ around the nucleus. The first shell is ________ to the nucleus and can hold a maximum of ________ electrons. The second and third shells can each hold a maximum of ________ electrons.

2 Copy and complete the table showing the electron arrangement of the first twenty elements. Use the diagrams on page 70 to help you.

| Element number | Atom of | Electronic structure |
|---|---|---|
| 1 | hydrogen | 1 |
| 2 | helium | 2 |
| 3 | lithium | 2, 1 |
| 4 | beryllium | |
| 5 | | |
| 6 | | |
| 7 | | |
| 8 | | |
| 9 | | |
| 10 | neon | |
| 11 | | |
| 12 | | |
| 13 | | |
| 14 | | |
| 15 | | |
| 16 | | |
| 17 | | |
| 18 | | |
| 19 | | |
| 20 | calcium | 2, 8, 8, 2 |

3 Is each of the following statements true or false?

A The first shell can contain a maximum of two electrons.

B Electrons are in the nucleus of an atom.

C The number of electrons in an atom is the same as the number of protons.

D In a calcium atom electrons are in three electron shells.

### Key ideas

- *Electrons go into shells around the nucleus.*
- *The first shell is closest to the nucleus and holds a maximum of two electrons.*
- *The second and third shells hold a maximum of eight electrons each.*
- *Electrons go into the first shell until it is full, then the second, then the third and so on.*

# 3:4 Ordering the elements

## Looking for patterns

**a What pattern can you see in this jumble of black and white?**

When you first turned this page, the black and white picture probably appeared as a complete jumble. It is, in fact, Dalmatians in the snow, and now that you know that, it probably looks obvious!

An even worse problem faced the early chemists as they searched for patterns in the huge amount of information they were uncovering about the different elements. They did not know how many elements there were, so they did not know how much information was still missing. They had no idea why elements were different.

It is only thanks to their pioneering work that you are now in the fortunate position of knowing what to look for!

## Ordering the elements

Gold, copper, sulphur and mercury had been known for thousands of years. By the year1800, thirty elements were known.

Many elements were discovered in the next 50 years. With so much new information, chemists wanted to find some pattern.

Some chemists found that elements could be grouped in families. For example, the reactive non-metals chlorine, bromine and iodine were such a family.

It was only when it became possible to arrange the elements in order of increasing **atomic mass** that real progress was made. Once this was achieved, a pattern started to emerge.

Johann Dobereiner picked out groups of three elements with similar properties. He called these **triads**. He noticed that the atomic mass of the middle element was the average of the other two.

For example:

| | |
|---|---|
| Calcium | 40 |
| Strontium | 88 |
| Barium | 137 |

The table shows the first twenty elements.

**b Pick out three similar elements from this list. They make another triad.**

**lithium silicon sodium phosphorus potassium**

**c What pattern can you see in this list of elements?**

| $Z$ | $A_r$ | Element | | Properties |
|---|---|---|---|---|
| 1 | 1 | hydrogen | H | very reactive gas |
| 2 | 4 | helium | He | unreactive gas |
| 3 | 7 | lithium | Li | very reactive metal |
| 4 | 9 | beryllium | Be | reactive metal |
| 5 | 11 | boron | B | non-metal |
| 6 | 12 | carbon | C | non-metal |
| 7 | 14 | nitrogen | N | non-metal |
| 8 | 16 | oxygen | O | reactive gas |
| 9 | 19 | fluorine | F | very reactive gas |
| 10 | 20 | neon | Ne | unreactive gas |
| 11 | 23 | sodium | Na | very reactive metal |
| 12 | 24 | magnesium | Mg | reactive metal |
| 13 | 27 | aluminium | Al | metal |
| 14 | 28 | silicon | Si | non-metal |
| 15 | 31 | phosphorus | P | non-metal |
| 16 | 32 | sulphur | S | reactive non-metal |
| 17 | 35.5 | chlorine | Cl | very reactive gas |
| 18 | 40 | argon | Ar | unreactive gas |
| 19 | 39 | potassium | K | very reactive metal |
| 20 | 40 | calcium | Ca | reactive metal |

**The first 20 elements: Newlands did not know about helium, neon and argon so his list of 17 elements repeated exactly like a musical scale.**

## The first 20 elements

When the first 20 elements are arranged in order of increasing atomic mass, they show a repeating pattern. This was first spotted by the British chemist John Newlands in 1864. He compared this to the repetition of notes on a musical scale and called it the 'law of octaves'.

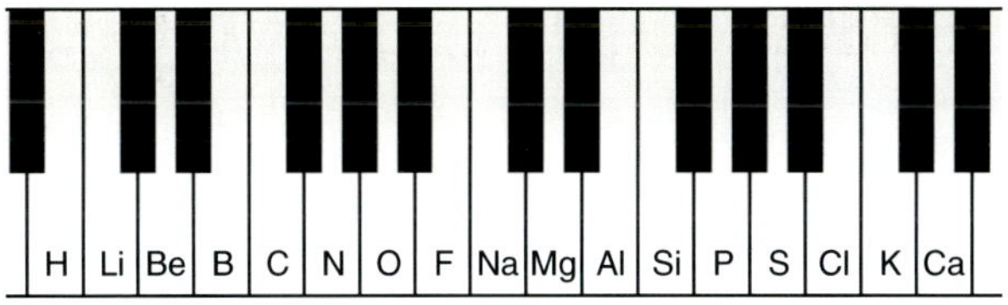


Newlands' scale.

**d Which elements were missing from Newlands' scale as shown here?**

Dmitri Mendeléev.

**Did you know?**
Mendeléev used Russian in his writing. We can still pick out the symbols for the elements in his writings.

## Mendeléev's table

This simple pattern breaks down after the first twenty elements. Newlands' ideas were not taken up immediately by other scientists. Just five years after Newlands, in 1869, the Russian chemist Dmitri Mendeléev looked for patterns in the elements. He arranged elements in order of increasing atomic mass but put elements with similar properties in the same vertical column.

Mendeléev's periodic table had its problems. He had had to put iodine out of position compared to its relative atomic mass in order to keep it in the same group as chlorine. Many scientists did not like this.

**e The stamp shown gives the day, month and year of the first periodic table. What are these?**

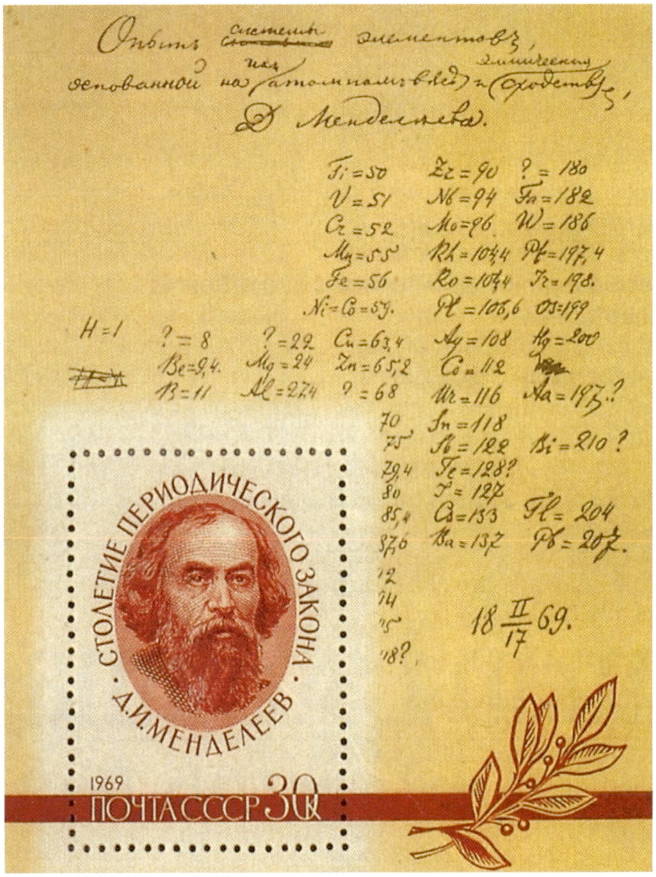


Celebration stamp produced for the 100th anniversary of the first periodic table.

Mendeléev had realised that some elements were missing. They were undiscovered at this time. He left gaps in his table. He was so confident that he even used his table to predict the properties of these 'unknowns'. When some of the missing elements were discovered, the accuracy of his predictions gave great support to his ideas.

**f What did Mendeléev use in his table for elements that had not been discovered?**

## Questions

1 Three chemists who helped to put elements into order were Mendeléev, Dobereiner and Newlands. Put them in the correct date order.

2 Copy and complete the sentences. Use words from the list.

atomic mass elements gaps row

Mendeléev arranged the known _______ in order of increasing _______. In his table elements with similar properties were in the same _______. He left _______ for elements that had not been discovered and made predictions about these elements.

## Key ideas

- *If the elements are arranged in order of atomic mass, a repeating pattern of properties can be seen.*
- *Mendeléev produced the first periodic table. He left gaps for elements that had not yet been discovered.*

# 3:5 The modern periodic table

## Introducing the periodic table

A modern table of the first 20 elements arranged like this clearly shows the repeating (or periodic) pattern, and so is called the **periodic table** of the elements.

The elements are arranged according to their atomic number (proton number). Ordered in this way, iodine and potassium are clearly in the correct position.

A copy of the full modern periodic table can be found at the back of this book.

**a How many elements are shown in the modern periodic table?**

Only 92 elements occur naturally. Scientists working at an American university have made the other elements.

**b Where is this university? (Hint: The element with atomic number 98 might help you.)**

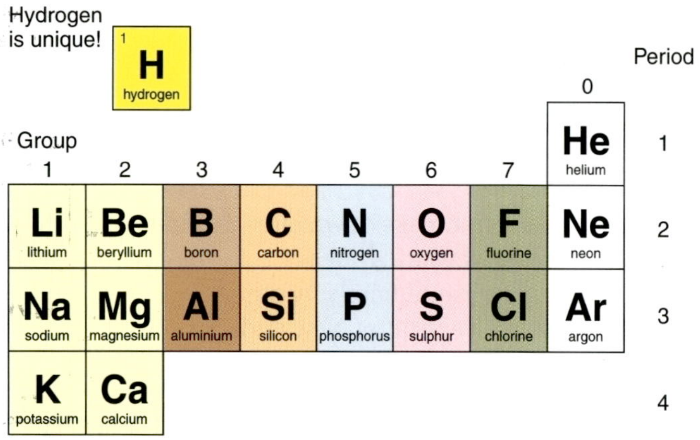


**A simple periodic table for the first 20 elements.**

The modern periodic table can be divided into different parts. These include:

**1 The main block of the periodic table**
The elements shown in blue make up the main block of the periodic table.

The numbered rows across the table are called **periods**. They are numbered 1–7.

The main vertical columns are called **groups**. They are numbered from 1 to 7 and then 0.

The bold line divides metals on the left-hand side from non-metals on the right-hand side.

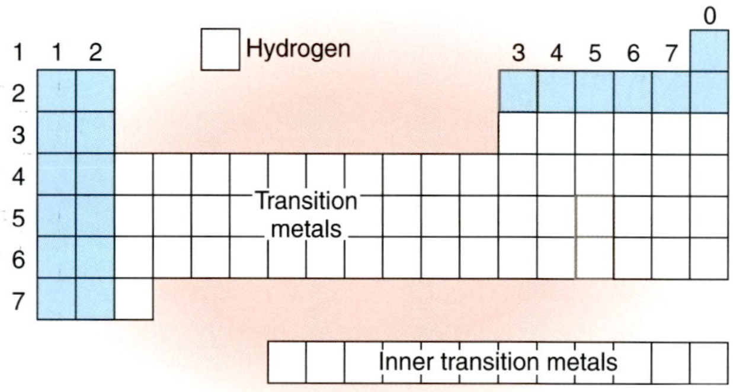


**c Are there more metals or more non-metals?**

Elements in the same family are in the same vertical column or group.

**d Pick out of this list an element similar to lithium, sodium and potassium.**

**magnesium   phosphorus   rubidium   sulphur**

**e Name a metal found in group 4.**

**f Are each of the following elements metals or non-metals?**

**aluminium   argon   beryllium   boron**
**calcium   phosphorus   sulphur**

You will need to recognise these families for your GCSE examination:

- alkali metals (Group 1), Module 9;
- halogens (Group 7), Module 3;
- noble gases (Group 0), Module 9.

**2 Hydrogen**
You will notice that hydrogen is on its own. Hydrogen has different properties from all the other elements.

**3 Transition metals**
This is the block of 37 metallic elements between groups 2 and 3. Iron and nickel are transition metals. Some properties of transition metals will be studied in Module 9.

## Questions

**1** Copy and complete the following sentences. Use words from this list.

**group periodic transition**

Elements are arranged in order of increasing atomic number in the modern _______ table. Elements in the same family are in the same vertical column or _______.

Elements between groups 2 and 3 are called _______ metals.

**2** Here is a list of elements.

**lithium beryllium boron carbon nitrogen oxygen fluorine neon**

a Which element is in group 3?

b Which element is in the same group as sodium?

c Which element is in group 7?

d These elements are in the same period. (True or False?)

e How many metals are there in the list?

**3** How many metals and how many non-metals are there in the first 20 elements? How are they arranged?

**4** The elements that are gases at room temperature and pressure are shown in the box.

a How many elements are gases at room temperature and atmospheric pressure?

b Where in the periodic table are the elements that are gases at room temperature and pressure?

| |
|---|
| Argon |
| Helium |
| Neon |
| Radon |
| Krypton |
| Xenon |
| Fluorine |
| Chlorine |
| Oxygen |
| Nitrogen |
| Hydrogen |

### Key ideas

- *Elements in the modern periodic table are arranged in order of increasing atomic number.*
- *Elements with similar properties are in the same group.*
- *The bold line divides metals on the left-hand side from non-metals on the right-hand side.*

# 3:6 Patterns in the periodic table

## Patterns in the periodic table

There are many patterns in the periodic table. They occur because there is a pattern in the arrangement of electrons across each period and down each group.

This table shows part of the periodic table.

| | 1 | 2 | 3 | 4 | 5 | 6 | 7 | 8 |
|---|---|---|---|---|---|---|---|---|
| 1 | | | | | | | | He |
| 2 | Li | Be | B | C | N | O | F | Ne |
| 3 | Na | Mg | Al | Si | P | S | Cl | Ar |
| 4 | K | Ca | Ga | Ge | As | Se | Br | Kr |
| 5 | Rb | Sr | In | Sn | Sb | Te | I | Xe |

## The electron link

The pattern of the periodic table links in precisely with the way the electrons are arranged in the atoms. This was, of course, unknown to Mendeléev.

The period number tells you how many shells containing electrons the element has. The group number tells you how many electrons an element has in its outer shell.

For example, all the elements in period 2 have two shells.

**a** **Carbon is in period 2. How many electron shells are there in a carbon atom?**

**b** **How many shells of electrons are there in a calcium atom?**

The number of electrons in the outer shell is the same as the group number.

Lithium is in group 1, so it only has one electron in its outer shell.

**c** **Carbon is in group 4. How many electrons are there in the outer shell of a carbon atom?**

**d** **How many electrons does iodine (I) have in its outer shell?**

**e** **Which element has two electrons in its fifth shell?**

## Patterns in reactivity

Down any group there is a pattern in physical and chemical **properties**. This means that, although the elements are not identical, if you know the properties of some of the elements in a group, you can predict the properties of others.

## Patterns in the formulae of compounds

There are patterns in the formulae of compounds. Understanding the patterns may help you to write correct formulae.

This table shows the formulae of the chlorides of some of the elements in the main block of the periodic table.

You should be able to see a pattern in these formulae.

Elements in the same group form chlorides with the same formula, e.g. LiCl and NaCl.

The number of chlorine atoms combined with one of the atoms of the element is the same as the group number or the group number subtracted from eight.

| Group | Period 2 | Period 3 | Period 4 |
|---|---|---|---|
| 1 | LiCl | NaCl | KCl |
| 2 | $BeCl_2$ | $MgCl_2$ | $CaCl_2$ |
| 3 | $BCl_3$ | $AlCl_3$ | $GaCl_3$ |
| 4 | $CCl_4$ | $SiCl_4$ | $GeCl_4$ |
| 5 | $NCl_3$ | $PCl_3$ | $AsCl_3$ |
| 6 | $OCl_2$ | $SCl_2$ | $SeCl_2$ |
| 7 | FCl | $Cl_2$ | BrCl |
| 0 | no oxides | no oxides | no oxides |

For example:

- $AlCl_3$ – aluminium, which is in group 3, combines with three chlorine atoms;
- $PCl_3$ – phosphorus, which is in group 5, combines with 3 (8 – 5) chlorine atoms.

**f** Caesium (Cs) is in group 1. Look at the formulae of lithium chloride, sodium chloride and potassium chloride. Predict the formula of caesium chloride.

**g** Tellurium (Te) is in group 6. Look at the formulae of sulphur chloride and selenium chloride. Predict the formula of tellurium chloride.

You should see a similar pattern with oxides.

**h** Write the formulae for barium oxide. Look at the periodic table to find out which group barium is in.

The periodic table is useful in making predictions about elements from the position of the element.

| Group | Period 2 | Period 3 | Period 4 |
|---|---|---|---|
| 1 | $Li_2O$ | $Na_2O$ | $K_2O$ |
| 2 | BeO | MgO | CaO |
| 3 | $B_2O_3$ | $Al_2O_3$ | $Ga_2O_3$ |
| 4 | $CO_2$ | $SiO_2$ | $GeO_2$ |
| 5 | $N_2O_3$ | $P_2O_3$ | $As_2O_3$ |
| 6 | – | $SO_2$ | $SeO_2$ |
| 7 | $F_2O$ | $Cl_2O$ | $Br_2O$ |
| 0 | no oxides | no oxides | no oxides |

| Group | Period 2 | Period 3 | Period 4 |
|---|---|---|---|
| 1 | LiH | | KH |
| 2 | $BeH_2$ | $MgH_2$ | |
| 3 | $BH_3$ | $AlH_3$ | $GaH_3$ |
| 4 | | $SiH_4$ | $GeH_4$ |
| 5 | $NH_3$ | | $AsH_3$ |
| 6 | – | $SH_2$ | |
| 7 | FH | | BrH |
| 0 | no hydrides | no hydrides | no hydrides |

## Questions

**1** Copy and complete the sentences. Use words from this list.

**group** **period** **properties**

The number of shells of electrons is the same as the ________ number.

The number of electrons in the outer shell is the same as the ________ number.

There is a pattern in the ________ of elements down each group.

**2** Find the element boron (atomic number 5) in the periodic table.

**a** Is it a metal or a non-metal?

**b** How many electrons are there in a boron atom?

**c** How many shells of electrons are there?

**d** How many electrons are there in the outer shell?

**e** Write down the electronic structure of a boron atom.

**f** Which element in period 3 most closely resembles boron?

**3** The table above right shows the pattern of hydrides (compounds of an element with hydrogen) for some of the main block elements.

**a** Copy and complete the table.

**b** Use the periodic table to write the formulae of: i caesium hydride, ii barium hydride and iii iodine hydride.

**4** An element has electrons in five shells. There are four electrons in the outer shell. What is the name of this element?

### Key ideas

- *The periodic table has the elements arranged in groups.*
- *Elements in the same group have similar properties. There is a gradual change in properties of the elements down each group.*
- *There is a pattern in the electron arrangements of atoms of the different elements. The number of shells of electrons is the same as the period number. The number of electrons in the outer shell is the same as the group number.*
- *The periodic table can be used to predict the formulae of compounds.*
- *The periodic table is a powerful tool for predicting how elements react.*

# 3:7 The halogens

Group 7 of the periodic table contains a family of very reactive non-metals called the **halogens**. Four of these elements are:

- fluorine, F;
- chlorine, Cl;
- bromine, Br;
- iodine, I.

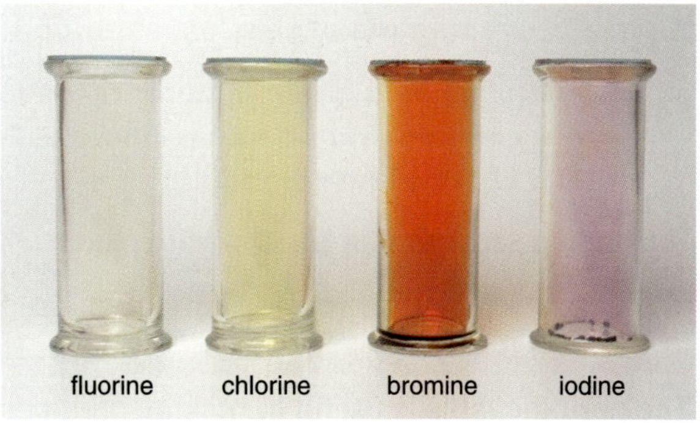


**The halogens form coloured gases.**

Like many other non-metals they:

- have low melting and boiling points;
- form soft crumbly solids (if cooled down enough);
- do not conduct electricity at all.

This table gives information about the halogens.

| Element | Atomic number | Melting point (°C) | Boiling point (°C) | Density at room temperature and atmospheric pressure ($g/dm^3$) | Appearance at room temperature and atmospheric pressure |
|---|---|---|---|---|---|
| fluorine | 9 | –220 | –188 | 1.6 | colourless gas |
| chlorine | 17 | –101 | –34 | 3.0 | greenish-yellow gas |
| bromine | 35 | –7 | 58 | 3.1 | dark red liquid |
| iodine | 53 | 114 | 183 | 4.9 | black shiny solid |

**a What happens to the melting points of these elements as the atomic number gets larger?**

**b What happens to the boiling points of these elements as the atomic number increases?**

**c What is the pattern in the densities of these elements with increasing atomic number?**

Although there are similarities in physical properties of the halogens down the group, there is a gradual change down the group.

All of the halogens have simple molecular structures with **diatomic** molecules, i.e. $F_2$, $Cl_2$, $Br_2$ and $I_2$.

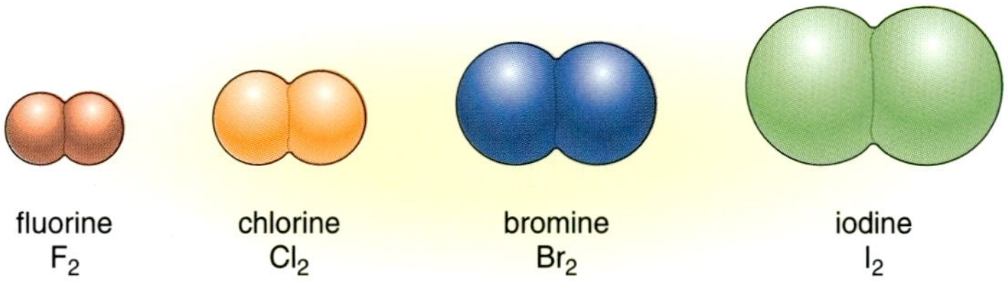


## How the halogens react

All of the halogens are reactive non-metals. Halogens react with most metals to form **salts**.

The halogen atoms are all just one electron short of a full outer shell. These atoms want to gain an extra electron each and this is what makes them reactive.

With metals, however, they can capture electrons to form single negative ions and so form ionic compounds such as sodium chloride, $Na^+Cl^-$.

**Did you know?**
The word halogen means salt producer.

Note how the halogen name changes as it forms an ion:

- chlori**ne** forms chlori**de** ($Cl^-$) ions;
- bromi**ne** forms bromi**de** ($Br^-$) ions;
- iodi**ne** form iodi**de** ($I^-$) ions.

**d** **What ion do you think forms from fluorine?**

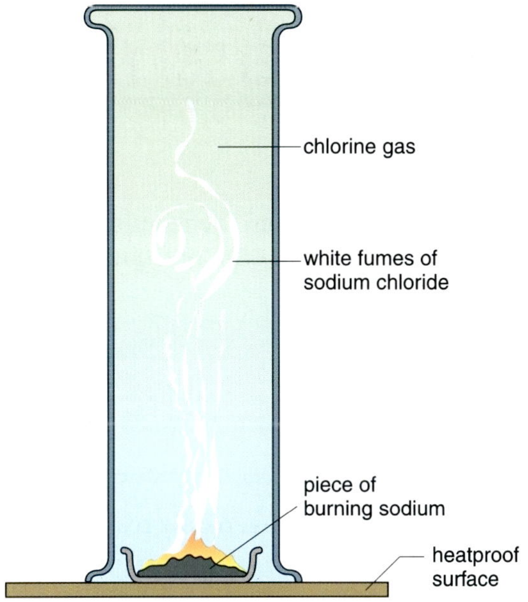

**Sodium burning in a gas jar of chlorine.**

### Reaction with sodium

Sodium is a very reactive metal.

A piece of sodium burns in chlorine gas to form sodium chloride.

sodium + chlorine → sodium chloride

Sodium chloride is often called common salt.

The other halogens react in a similar way.

**e** **What is the name of the compound formed when sodium burns in bromine?**

### Reaction with iron

If iron comes into contact with chlorine a rapid reaction takes place.

Iron reacts with chlorine to make iron(III) chloride.

iron + chlorine → iron(III) chloride

During the reaction the iron glows brightly.

If bromine comes into contact with iron, a reaction takes place. This reaction will be slower than the reaction of iron with chlorine.

Iron and iodine react very slowly.

**f** **What evidence is there that the reaction between iron and chlorine is exothermic?**

**g** **Use the information to put the three halogens, bromine, chlorine and iodine, in order of decreasing reactivity.**

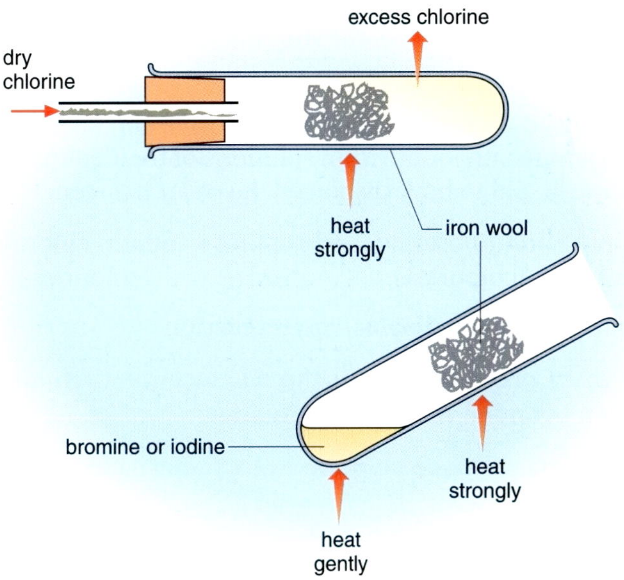

**Reactions of iron with chlorine, bromine and iodine.**

## Questions

1 a Which halogen is a colourless gas?

b Which halogen would be a solid at the North Pole and a gas at the tropics?

c Which halogen has the lowest boiling point?

d Which halogen is a low melting point solid?

2 Suggest why fluorine is not used in schools but chlorine, bromine and iodine are.

3 Fluoride is sometimes added to tap water. Why?

4 Astatine (atomic number 85) is a halogen element. It is below iodine in group 7. Predict the physical properties of astatine.

## Key ideas

- *Group 7 contains a family of very reactive non-metals with coloured vapours.*
- *As you go down the group, the elements change from gases, through liquids to solids.*
- *As you go down the group, the elements become less reactive.*

# 3:8 Reactivity of the halogens

You will already know from the reactions of halogens that chlorine is more reactive than bromine and bromine is more reactive than iodine.

## Confirming this pattern

The halogens become less reactive as you go down the group.

- Chlorine and hydrogen explode together in sunlight.
- Bromine and hydrogen only react if heated.
- Iodine and hydrogen only partially react even if heated.

The word equation for the reaction of hydrogen and chlorine is:

hydrogen + chlorine → hydrogen chloride

**a Write the word equation for the reaction of hydrogen and bromine.**

Because of this difference in reactivity, chlorine can take the place of bromine and iodine from solutions of their compounds. The chlorine is said to have **displaced** the other halogen.

| chlorine + | sodium bromide → | bromine + | sodium chloride |
|---|---|---|---|
| (green) | (colourless) | (brown) | (colourless) |

This is an example of a **displacement reaction**.

**b Write the word equation for the displacement reaction between bromine and sodium iodide.**

**c Could iodine displace chlorine from sodium chloride? Explain your answer.**

chlorine
bromine ↓ decreasing reactivity
iodine

## Uses of halogens

Iodine is used as an antiseptic. If you cut your finger, a solution of iodine in ethanol may be used to kill bacteria around the cut. The skin is stained brown.

Chlorine has a wide range of uses. Many of these depend upon chlorine being able to kill harmful bacteria.

**d Chlorine bleach bottles have a cap that is difficult for children to remove. Why?**

**e Why should bleach not be stored in old lemonade bottles?**

**Did you know?**

Fritz Haber (who developed the Haber process) was in charge of the development of chlorine and other poisonous gases for Germany in the First World War. In 1919, Haber was awarded a Nobel prize for his work with ammonia. Several scientists refused to accept their Nobel prizes as a protest against Haber and his work with poisonous gases.

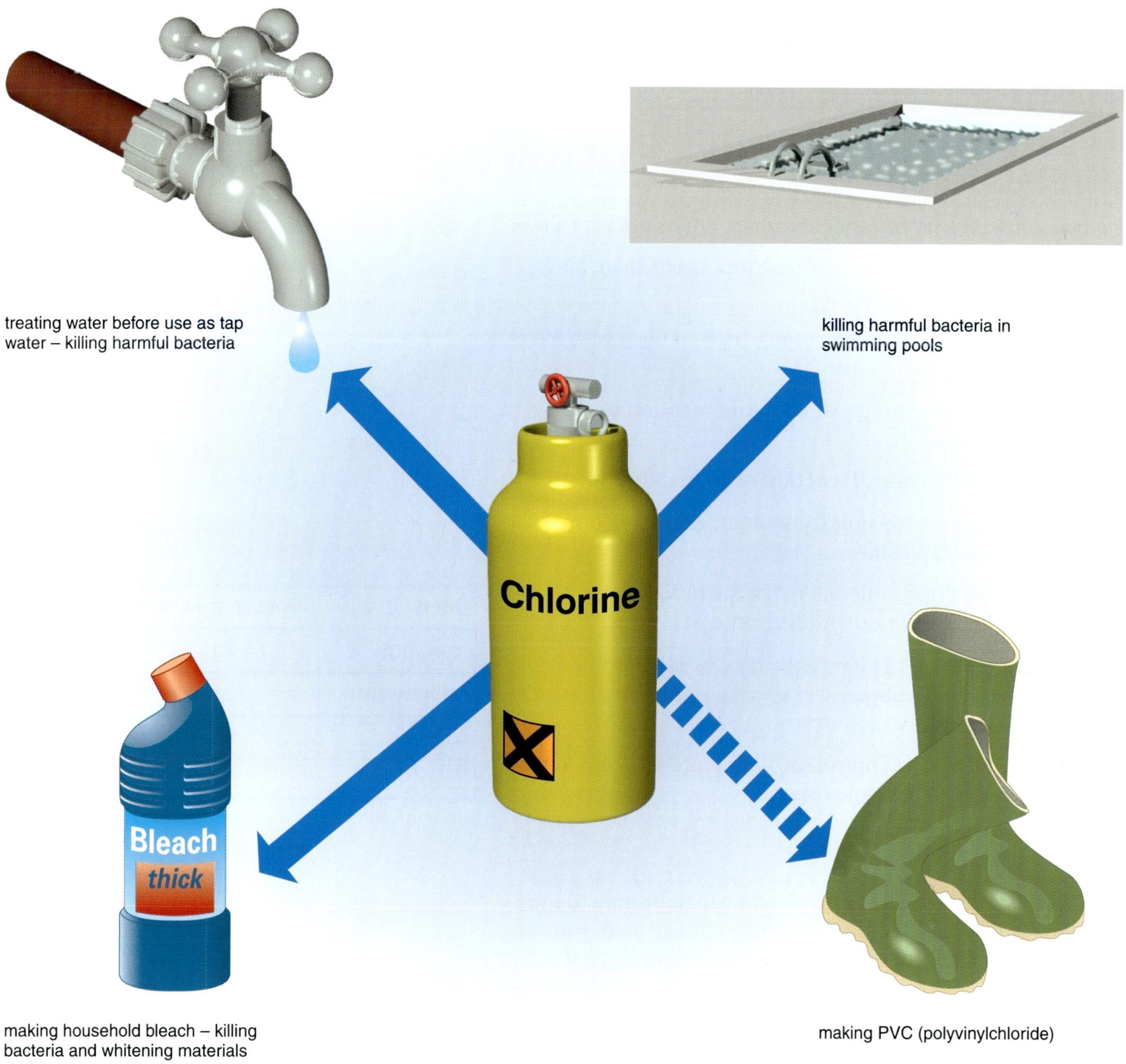


**Some uses of chlorine.**

## Questions

1 A few drops of orange–red bromine is added to a colourless solution of potassium iodide.

- a What colour change would you see?
- b What type of reaction is taking place?
- c Write a word equation for this reaction.
- d Explain why the reaction takes place.

2 Seawater contains bromide ions. Bromine is made by bubbling chlorine gas through seawater. Explain why this reaction works.

3 Write down three uses of chlorine that rely on the power of chlorine to kill harmful bacteria.

### Key ideas

- *Group 7 contains a family of very reactive non-metals with coloured vapours.*
- *Chlorine is used to kill bacteria. It is used in swimming pools, added to the water supply in small quantities and used to make household bleaches.*

# 3:9 Introducing rates of reaction

## Fast and slow

Some reactions are very fast indeed, whilst others are very slow.

- In a car engine, the petrol/air mixture reacts in an almost instantaneous explosion.
- The body of a car may take many years to react with air and water and turn to rust.

**a What provides the energy to start the reaction of the petrol/air mixture?**

It is very important to control **reaction rates** in industry.

- If a reaction runs too quickly, it might go out of control and cause an explosion.
- If it runs too slowly, this will make the process very inefficient and will raise production costs.

**b A chemical reaction occurs when a cake is baked in an oven. What would happen to the cake if the reaction went too fast?**

Reaction rates need to be controlled, but before you can do that you need to be able to measure them.

| Metal (0.1g) | Time to react (s) |
|---|---|
| calcium | 13 |
| magnesium | 24 |
| zinc | 90 |
| iron | 120 |

## Measuring reaction rates

To find how fast a reaction is going, you need to be able to time something. In the case of a metal reacting with an acid, you can time how long it takes for the metal to react and disappear completely.

You could set up an experiment, for example, where samples of different metals were reacted with a large amount of acid. You would need a large amount of acid to make sure that all of the metal would react. Each time you would need to use the same amounts of each metal and acid in order to make it a fair test.

The time is taken from when the metal is added to the acid until all the metal has reacted.

The longer it takes for the reaction to occur, the slower the reaction is. To find the average speed (or **rate**) of the reaction, you can divide the mass of metal lost by the time it took. This would give you a reaction rate in terms of grams per second (g/s).

**c Put the four metals in the table above in order of rate of reaction. Put the fastest reaction first.**

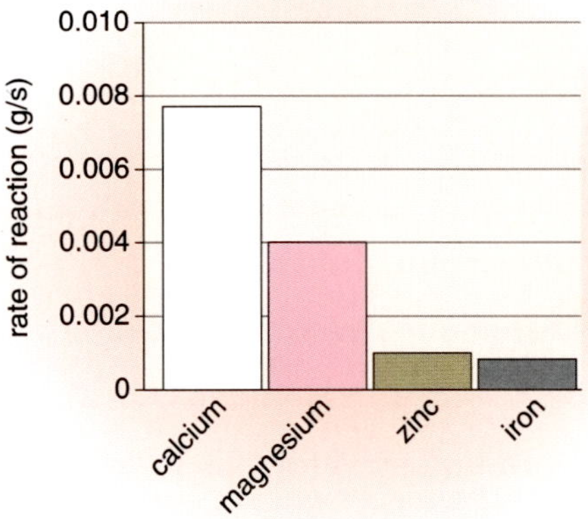


**Bar chart showing the rates of reaction of four metals with acid.**

## What can you measure?

To time a reaction, you need some way of following either the loss of the reacting chemicals (**reactants**) or the increase in the **products** of the reaction. In a simple metal/acid reaction:

| *reactants* | | *products* | |
|---|---|---|---|
| magnesium + | sulphuric acid → | magnesium sulphate | + hydrogen |

The simple method used above measured how long it took for one of the reactants (the metal) to be used up in the reaction. But this reaction also produces bubbles of hydrogen gas.

You can measure how fast this product is produced by using a **gas syringe**. The gas produced is collected in the gas syringe.

**d** How is the rate of reaction shown here changing with time?

**e** What is the rate of reaction when the graph is flat?

Alternatively, you can take readings every few seconds and plot a graph of the volume of hydrogen formed against the time. Graphs like this clearly show that the reaction slows down as the reactants get 'used up'.

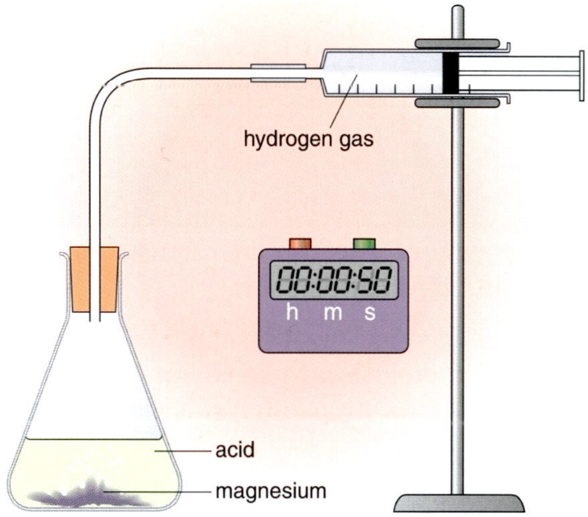

**Rate of reaction experiment with magnesium and sulphuric acid.**

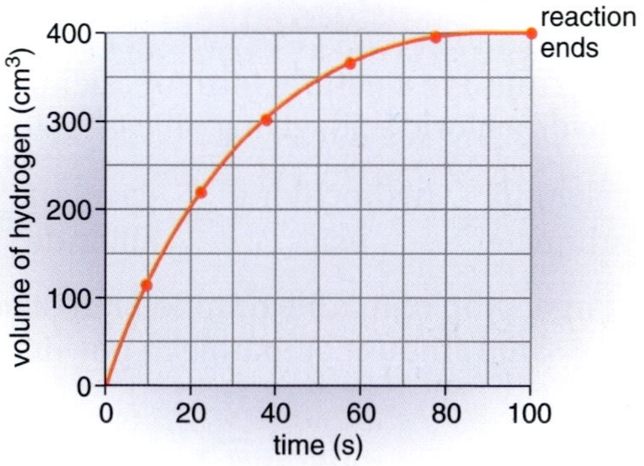

**Graph of volume of gas collected against time.**

## Questions

**1** Here are four chemical reactions. Put them in order of rate of reaction. Put the fastest one first.

**A** Rusting of iron

**B** Explosion of gas and air.

**C** Erosion of a gravestone.

**D** Reaction of an Alka Seltzer tablet with water.

**2** Limestone reacts with acid, producing carbon dioxide gas. This graph shows the volume of gas produced over time.

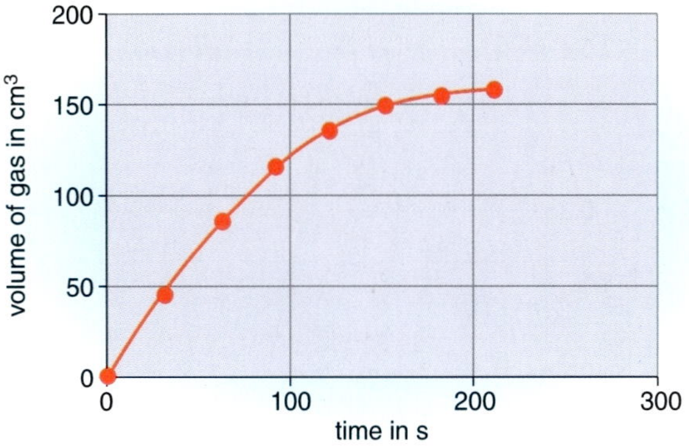


**a** After how many seconds is the reaction finished? How can you tell this?

**b** Is the reaction faster after 50 seconds, 100 seconds or 200 seconds? How can you tell this?

**c** What was the final volume of gas collected?

## Key ideas

- *Different reactions can run at very different rates. A fast reaction is completed in a short period of time. A slow reaction takes a long time to complete.*
- *In industry it is very important to control the rate of reaction.*
- *You can follow the rate of a reaction either by:*
  - *measuring how fast a reactant is used up, or*
  - *measuring how fast a product is formed.*

# 3:10 Speed up by getting together

Controlling the rate of a reaction is very important in industry. If you can speed up a reaction, you can usually keep the costs down. There are many ways of doing this.

Before a chemical reaction can take place, the particles of the reacting chemicals have to collide. Anything you do that makes the particles collide more often will speed up the reaction.

Two ways of doing this would be:

- make the pieces smaller. This increases the **surface area** of a solid;
- make the solution stronger or increase the **concentration** of the solution.

**Reaction of lump of marble (left) and powdered calcium carbonate (right) with hydrochloric acid.**

## Investigating surface area

Marble chips are a natural form of calcium carbonate. They react with hydrochloric acid, giving off carbon dioxide gas.

calcium carbonate + hydrochloric acid → calcium chloride + water + carbon dioxide

One large lump of marble dropped into a beaker fizzes steadily. But the same amount of marble in powdered form foams up out of the beaker.

**a What does the difference between the two reactions show about the rate of reaction?**

This is because the acid particles can only collide with the carbonate particles exposed at the surface. Breaking up a solid exposes more and more surface area and so makes the reaction go faster.

**b Why does chewing your food well help you to digest it?**

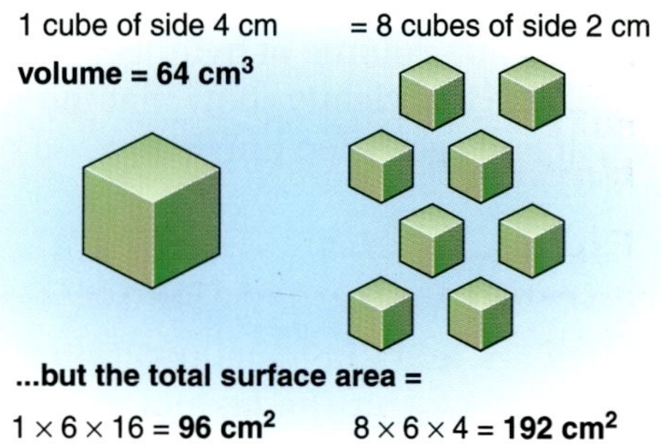


**Did you know?**

Mixtures of coal dust and air in a coal mine can explode but there is no such risk with lumps of coal.

In a flour mill care has to be taken not to have flour dust in the atmosphere. This can explode when mixed with air.

## Following the reaction

If this reaction takes place in an open beaker, the carbon dioxide escapes into the air. If this is done on a balance, the total mass will go down, as the carbon dioxide is lost. If you weigh the beaker and its contents at intervals you can compare the rates of the two reactions.

**c Which reaction is faster? How can you tell?**

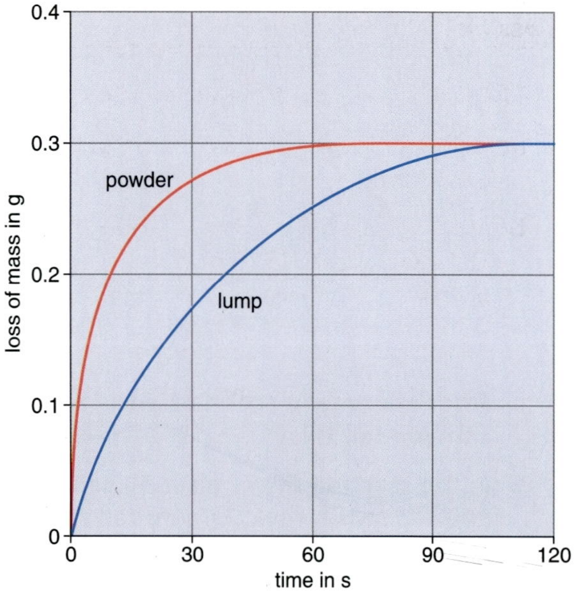


## Investigating concentration

The same reaction can be used to show how reaction rate changes when the concentration is changed. You can do this very simply by dropping marble chips into beakers of acid of increasing strength. You should not be surprised to find that stronger acid gives a faster reaction. What is the explanation?

**d Sketch the shape of the graph you would expect to get if you plotted rate of reaction against concentration.**

Remember that before two particles can react, they must meet! If the acid is at low concentrations, the acid particles will be widely spread out in the water, and the number of collisions between them and the marble chips will be limited. At higher concentrations, however, the chances of a collision between the acid particles and the marble chips are greatly increased.

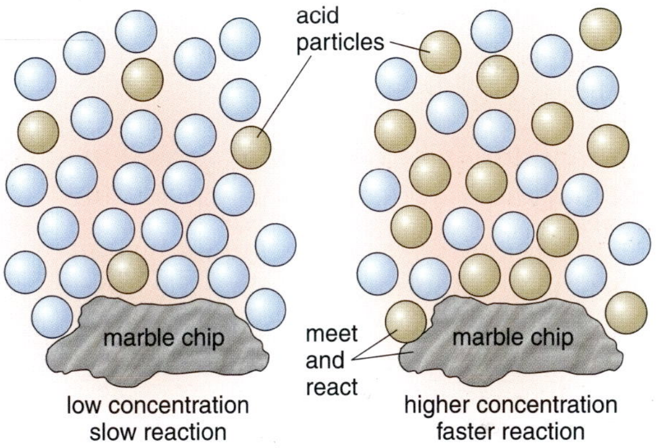


## Pressure, too

In reactions involving gases, the equivalent factor to this is pressure. All else being equal, the greater the pressure, the greater the number of particles in a given space. The closer the particles are together, the more chance they have of colliding with another. If you double the pressure, you will double the rate of reaction.

Graphs drawn for rate against concentration and rate against pressure look very similar.

## Questions

1 Why is it that delicate carvings on churches seem to suffer more from the effects of acid rain than large blocks of the same stone?

2 Explain as clearly as possible why breaking up a solid increases the surface area and so speeds up any chemical reaction.

3 Potatoes cook in boiling water in about 20 minutes. However, if the potatoes are cut up into small pieces they cook much faster. Explain why.

4 Suggest why dilute sulphuric acid spilt onto your clothes may not burn a hole at first – but will do if you leave it to 'dry' (let the water evaporate).

5 Here is the raw data from a marble/acid experiment.

| Time (s) | 0 | 20 | 40 | 60 | 80 | 100 |
|---|---|---|---|---|---|---|
| Mass (g) | 323.97 | 323.35 | 323.03 | 322.88 | 322.80 | 322.76 |
| Total mass loss (g) | 0.00 | | | | | |

The mass includes the beaker each time.

a Copy the table and work out the total mass loss each time.

b Plot a graph of this mass loss ($y$) against time ($x$). On your graph, sketch the lines you would expect if you:

i doubled the concentration;

ii halved the concentration.

## Key ideas

- *It is often important to be able to speed up reactions in industry.*
- *Particles must meet before they can react.*
- *Reactions can be speeded up by breaking up any solids and so increasing the surface area.*
- *Reactions can be speeded up by increasing the concentration of solutions or the pressure in gases.*

# 3:11 Heat up to speed up

If you think a reaction is going too slowly in the school laboratory, the first thing you would probably do would be to heat it up. It is a common experience in chemistry that heating makes things happen faster. Meat cooks faster in a hotter oven, but you keep milk cool in the fridge to stop it going 'off'.

What might surprise you is just how much an increase in temperature affects the rate of reaction. The 'rule of thumb' for most reactions is that the rate will double for every 10 °C or so rise in temperature.

**a These bottles of milk are the same age. Which do you think was left in a warm room, and which was left in the fridge?**

Which bottle of milk would you like to drink?

## Following a reaction

Sodium thiosulphate is a chemical used in photography. If you mix a clear solution of sodium thiosulphate with hydrochloric acid, a reaction occurs and a creamy-yellow **precipitate** of sulphur is produced.

sodium thiosulphate + hydrochloric acid → sodium chloride + sulphur dioxide + sulphur + water

The precipitate does not start to form straight away. It begins to appear after a short time, turning the mixture cloudy. You could time how long it takes for a pencil cross viewed through the liquid to disappear from view.

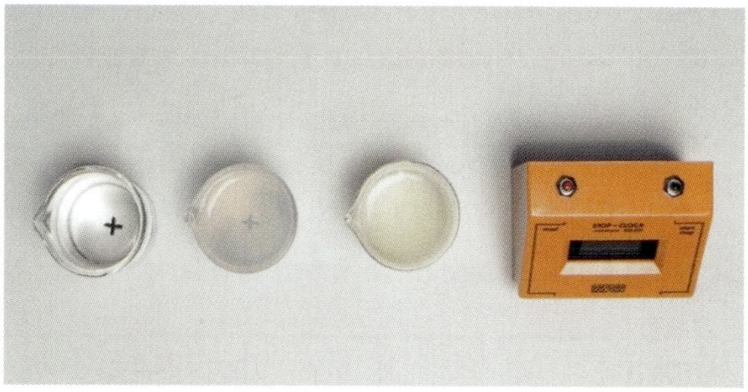

Going, going, gone! The thiosulphate/acid reaction turns the solution cloudy.

## Seeing the pattern

This experiment is easy to follow at different temperatures. All you need to do is to heat the thiosulphate solution to a higher temperature each time before adding a fixed volume of acid. You can measure the reaction temperature with a thermometer once the reactants have been mixed and the stopclock has been started.

If you plot the time taken for the cross to disappear against the temperature, you can clearly see how increasing the temperature speeds up the reaction.

**b What effect does raising the temperature have on: i) the time of the reaction and ii) the rate of reaction?**

**c From this graph, approximately what rise in temperature doubles the reaction rate? (Over what temperature range does the reaction time halve?)**

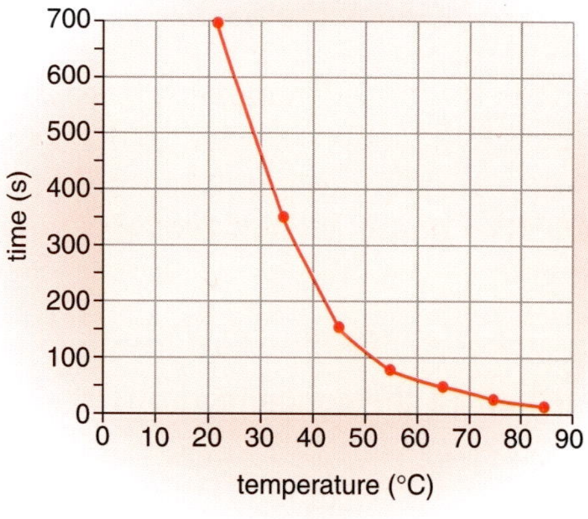

## A rate/temperature graph

Looking at a rate/temperature graph you might expect all industrial reactions to be run at as high a temperature as possible. High temperatures are often used but there are limits as the products may themselves break down if heated too much.

Cost is also important. You need a lot of energy to reach very high temperatures, and energy costs money. Industrial chemists are always on the lookout for the most cost-effective ways to run their chemical reactions.

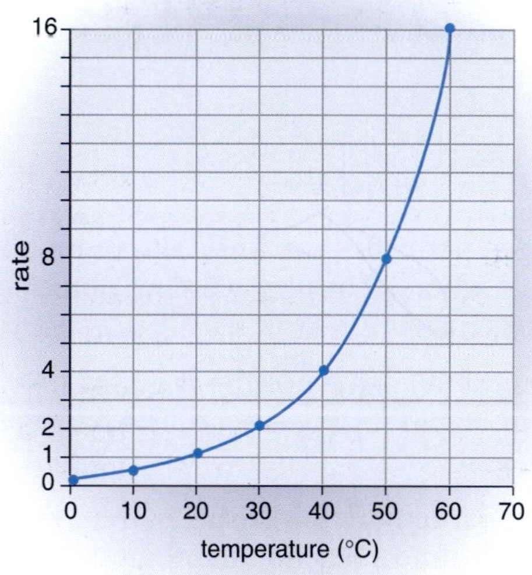


**Graph showing how rate of reaction changes with temperature.**

## What causes this temperature effect?

To understand why this happens, you need to think about what is going on at particle level. As you have seen, the reactant particles must collide, otherwise they cannot possibly 'change partners' in a chemical reaction. But that on its own is not enough – if it were, most chemical reactions would be almost instantaneous. They must also collide with enough energy to break the existing bonds.

**d Why doesn't a Bunsen burner light as soon as you turn on the gas?**

Now, at any given temperature, **collisions** of many different types will occur, from the maximum energy 'head-on' collisions to gentler glancing blows. But temperature will also have an effect, as the higher the temperature, the higher the average speed of the particles.

Clearly, the faster the particle, the more kinetic energy it has and so the more chance it has of having sufficient energy to react. If the temperature is raised, the average speed of the particles increases, so a larger proportion of the particles have a chance to react.

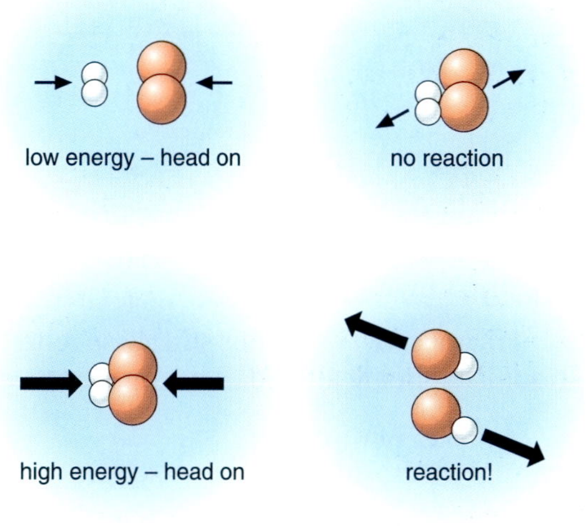


**You need the right kind of collision before you get a reaction.**

## Questions

1 **Superglue sets owing to a chemical reaction. Where will superglue set fastest, in the fridge, at room temperature or in a warm oven? Why?**

2 **Potatoes cook when heated in boiling water for about 20 minutes. They fry much faster in hot fat. Why is this?**

3 **Photographers use 'developer' solution to bring out the image in their photographs. If the film is left in too long, the photos will be very dark. If a film takes 5 minutes in solution at room temperature, what would happen if it were left in solution made with hot water for 5 minutes?**

### Key ideas

- *The rate of a chemical reaction increases if the temperature increases.*
- *Particles must collide with enough energy to break their bonds if they are to react.*
- *Raising the temperature means that more particles have sufficient energy to react.*

# 3:12 Catalysts and enzymes

## Everlasting activity!

Hydrogen peroxide is an unstable compound of hydrogen and oxygen. Left on its own, it will slowly break down into water and oxygen gas.

Manganese(IV) oxide powder is added to hydrogen peroxide solution. Rapid fizzing is seen. The fizzing is caused by oxygen gas.

Nothing particularly surprising in that, you might think – just another chemical reaction in progress. But the manganese(IV) oxide has not been changed.

If you filter the mixture after the reaction you can get the same amount of manganese(IV) oxide back – and use it again and again!

**a Use these diagrams to explain how you could show that manganese(IV) oxide is not taking part directly in this reaction.**

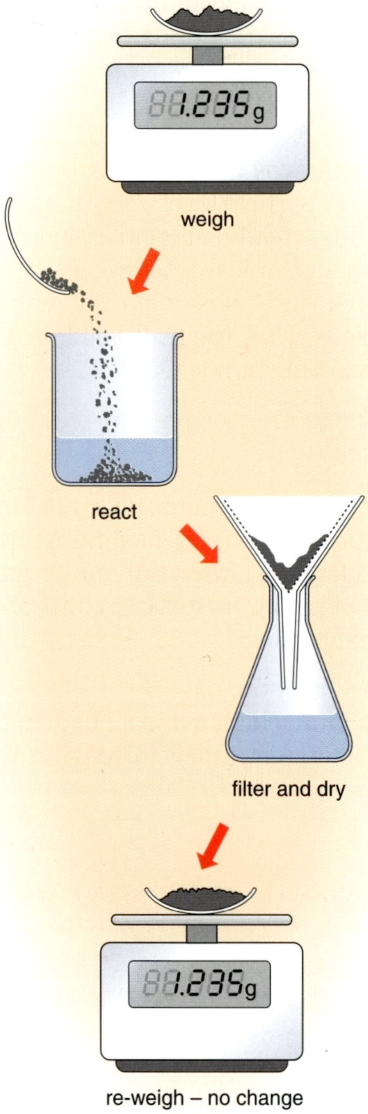


**Manganese(IV) oxide speeds up the breakdown of hydrogen peroxide, but remains unchanged itself.**

## What is happening?

The reaction that occurs is the simple breakdown of hydrogen peroxide to water and oxygen, which would have occurred slowly on its own.

The manganese(IV) oxide has simply speeded up this reaction without itself being altered. It has acted as a **catalyst**.

$$\text{hydrogen peroxide} \xrightarrow{MnO_2 \text{ catalyst}} \text{water} + \text{oxygen}$$

## Enzymes

**Chopped liver catalysis in action!**

If you added a drop of blood or some chopped liver to hydrogen peroxide the effect is even more dramatic than adding manganese(IV) oxide.

This is because of the presence of an organic, protein-based catalyst called catalase. These organic catalysts are called **enzymes**, and their action can be very powerful. Most life processes, such as digestion and respiration, depend on enzyme action.

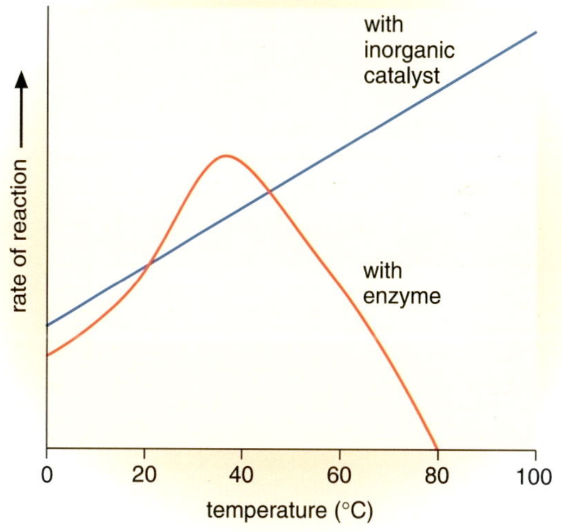


**Enzyme action peaks at 37 °C – that's body temperature.**

## So what's different?

With a catalyst, the rate of reaction rises further as the temperature increases, doubling every 10 °C or so. From 0 °C up to about 40 °C, enzymes behave in a similar way. But their action starts to tail off rapidly by 45 °C and, by about 60 °C, stops altogether.

**b In what ways are enzymes similar to inorganic catalysts? In what ways are they different?**

This is because enzymes are proteins like those in food and proteins break down at these higher temperatures. The enzyme is said to be **denatured.** It should not surprise you that the best temperature for enzyme action corresponds closely to your body temperature.

**c Why is it that enzyme action drops away so rapidly above 45 °C?**

Enzymes are also affected by pH. The enzymes that help to digest the food in your stomach work best at very low pH – this helps to explain why you stomach is full of hydrochloric acid. Other enzymes work best at different pH levels.

## Questions

**1** Complete the following sentences. Use words from the list.

**catalyst    denatured    enzymes**

A chemical that speeds up a reaction without being used up is called a ________. Some catalysts are proteins and these are called ________. They do not work at very high temperatures because the proteins are broken down or ________.

**2** Chemicals such as magnesium destroy hydrogen peroxide by reacting with the 'extra' oxygen:

hydrogen peroxide + magnesium → water + magnesium oxide

How is this different from catalysis?

**3** Many processes in the body involve enzymes. Suggest a reason why your body temperature is maintained at 37 °C.

## Key ideas

- *Catalysts speed up chemical reactions without being used up themselves.*
- *Enzymes are biological catalysts that work best at warm rather than hot conditions.*

# End of module questions

## Low demand

**Questions 1–3**

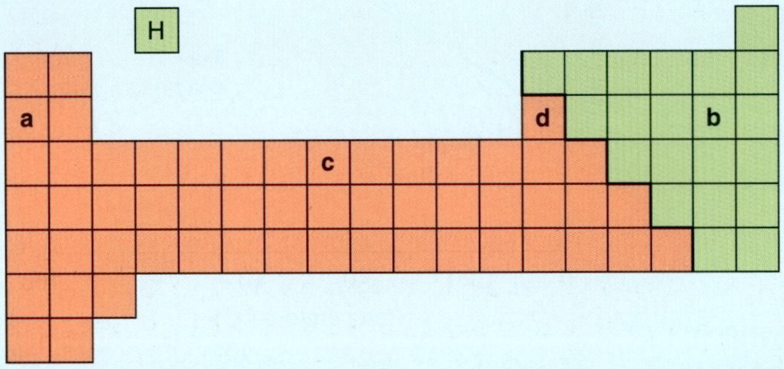


**1** Which letter shows the position of an alkali metal in the periodic table?

**A** a

**B** b

**C** c

**D** d

**2** Which letter shows the position of an element in period 4?

**A** a

**B** b

**C** c

**D** d

**3** Which letter shows the position of a non-metal?

**A** a

**B** b

**C** c

**D** d

**4** The periodic table is an arrangement of:

**A** elements

**B** compounds

**C** mixtures

**D** atoms.

**5** Which row is correct?

| | Particles in the nucleus | Particles in shells |
|---|---|---|
| A | electrons and protons | neutrons |
| B | protons and neutrons | electrons |
| C | electrons and neutrons | protons |
| D | protons | electrons and neutrons |

**6** Some marble chips and hydrochloric acid are reacting together in a flask.

Which of the following would speed up the reaction?

**A** Adding some water.

**B** Using powdered marble instead of marble chips.

**C** Cooling the reaction down.

**D** Putting a bung into the neck of the flask.

**7** A substance that speeds up a reaction without being used up is called:

**A** a catalyst

**B** a reactant

**C** a product

**D** a solid.

**8** The elements in the same group of the periodic table:

**A** have the same number of electrons in the outer shell

**B** have the same physical properties

**C** are in the same state

**D** have the same atomic number.

**9** Which of these elements is a halogen?

**A** sodium

**B** copper

**C** bromine

**D** sulphur

**10** Sodium burns in chlorine to form:

**A** sodium chloride

**B** sodium chlorate

**C** sodium oxide

**D** sodium chromate.

**11** Iodine solution is used on cuts in the skin. It is used to:

**A** keep the cut cool

**B** kill bacteria

**C** speed up the healing

**D** stop the cut bleeding.

**12** The building blocks of all materials are called:

**A** elements

**B** compounds

**C** ions

**D** mixtures.

## Standard demand

**13** A magnesium atom has an electron arrangement of 2, 8, 2. Which of the following will be the electron arrangement for an element in the same group as magnesium?

**A** 2, 8, 4

**B** 2, 8, 8, 2

**C** 2, 6

**D** 2, 8, 8, 1

**14** Potassium and bromine react together. Which of the following is the correct equation for this reaction?

**A** $K_2 + Br_2 \rightarrow 2KBr$

**B** $K + Br \rightarrow KBr$

**C** $2K + Br_2 \rightarrow 2KBr$

**D** $2K + 2Br \rightarrow (KBr)_2$

**Questions 15–17**

Hydrogen peroxide decomposes into water and oxygen gas. Manganese(IV) oxide is a catalyst for this reaction.

**15** In an experiment the volume of gas collected was measured at intervals. Which graph (A, B, C or D) would be obtained?

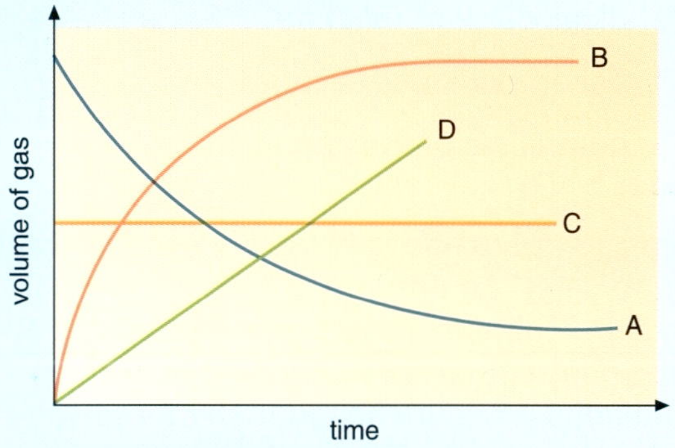


**16** In the same experiment, which graph (A, B, C or D) would show how the mass of manganese(IV) oxide changed during the experiment?

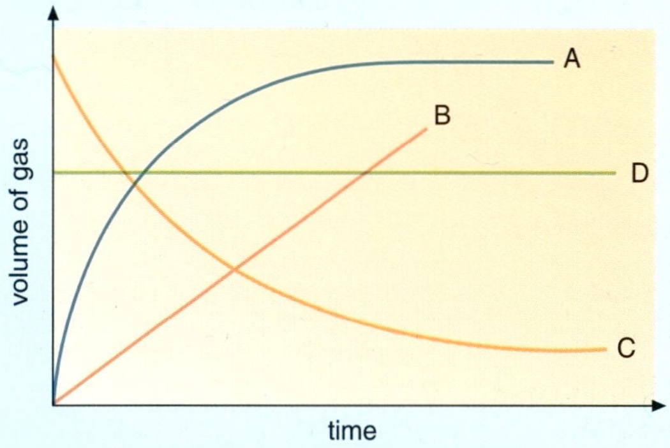


**17** The experiment slows down because:

**A** the mixture cools down

**B** the manganese(IV) oxide is used up

**C** there are fewer collisions by hydrogen peroxide particles

**D** water produced dilutes the solution.

**18** Chlorine is more reactive than bromine. Bromine is more reactive than iodine. Which of the following pairs react together?

**A** potassium chloride solution and bromine

**B** potassium chloride solution and iodine

**C** potassium bromide solution and iodine

**D** potassium bromide solution and chlorine

**19** The reaction in question 18 is a:

- **A** combustion reaction
- **B** displacement reaction
- **C** decomposition reaction
- **D** neutralisation reaction.

**20** Hydrogen peroxide is broken down into water and oxygen. An enzyme in celery speeds up the reaction. The experiment was carried out at three different temperatures but the other conditions were unchanged. The three temperatures were 20 °C, 40 °C and 80 °C.

Put the three reactions in order of increasing rate of reaction.

| | | | |
|---|---|---|---|
| **A** | 20 °C, | 40 °C, | 80 °C |
| **B** | 80 °C, | 40 °C , | 20 °C |
| **C** | 20 °C , | 40 °C, | 80 °C |
| **D** | 80 °C, | 20 °C, | 40 °C |

**21** Helium and argon:

- **A** are in the same period in the periodic table
- **B** have the same number of electrons in the outer shell
- **C** are in the same group in the periodic table
- **D** are called halogens.

**22** An element has atoms each containing 20 electrons.

These electrons will be found in:

- **A** one shell
- **B** two shells
- **C** three shells
- **D** four shells.

**23** With increasing atomic number the halogen elements have:

- **A** decreasing melting points
- **B** increasing boiling points
- **C** increasing reactivity
- **D** less colour.

**24** Which of the following electron arrangements is for atoms of elements in group 3 of the periodic table?

- **A** 2, 1
- **B** 2, 3
- **C** 2, 8, 1
- **D** 2, 8, 5

# Module 4 – Chemistry in action

This module is divided into five parts.

1 Crude oil is an important natural resource. After refining, various useful fractions are made. These have a wide range of uses.

2 Most of the uses of fractions from crude oil involve burning. The products of burning these fractions depend upon the burning conditions.

3 Some of the less valuable fractions can be converted into the raw materials used to make plastics. These have many uses in our modern world.

4 Enzymes are biological catalysts. They are used in a wide range of industries to produce useful products such as yoghurt and cheese.

5 Chemical reactions involve chemicals changing to form new products. There are different types of chemical reactions. Neutralisation and thermal decomposition are two types of chemical reaction.

Try these first.

1 When a substance burns it uses up a gas from the air. Which gas?

2 What is released to the surroundings when a substance burns?

3 What is another word for burning?

4 What method can be used to separate a mixture of liquids with different boiling points?

5 On a cold day a liquid drips out of a car exhaust when the car engine is started in the morning. What is this liquid?

6 What is the starting material used to make cheese and yoghurt?

7 What are the three elements combined in calcium carbonate?

8 Soils are often acidic. What type of substance is needed to neutralise acidic soils?

9 Calcium hydroxide is made up from three elements. What are these three elements?

10 A solution of calcium hydroxide is called limewater. What is limewater used for?

# 4:1 Crude oil

## There she blows!

Millions of gallons of **crude oil** gush from a hole in the ground.

It's time to celebrate if you have just drilled a well and made a lucky strike. This black liquid will make you rich. It can be used as a raw material to make:

- petrol and other fuels;
- plastics;
- paints and solvents;
- medicines.

**a But what about the environment? List three ways in which crude oil causes environmental problems.**

## How did crude oil form?

Crude oil is a **fossil fuel**. Millions of years ago, plants and animals in the sea died. They started to rot away.

High temperatures and high pressures acted on the remains. Without air or oxygen, the remains did not rot away completely and oil and natural gas were formed. They took millions of years to form deep in the Earth.

The oil and gas slowly moved up through the rocks. All the tiny droplets and bubbles collected together.

**b What conditions are needed for crude oil to form?**

**c A fuel is something we can get energy from. A fossil is the remains of a plant or animal found in the rock. Why are coal, oil and gas called fossil fuels?**

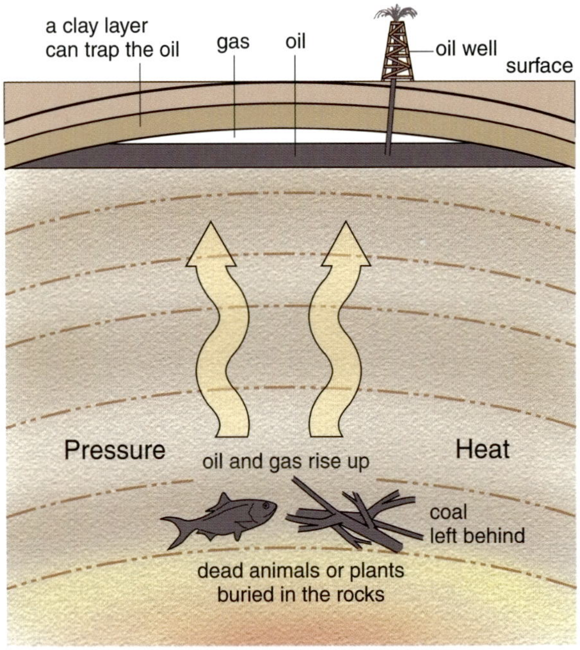


**Did you know?**
Crude oil has been known for thousands of years. It was known as pitch and the ancient Babylonians used it instead of cement to stick bricks together. Sir Walter Raleigh used pitch to keep his wooden ships watertight. In Texas, farmers would burn crude oil when they found it on the surface of the fields. It only became valuable when scientists found out how to turn it into useful products.

## What is crude oil made of?

Crude oil is a **mixture** of many different compounds. As they are not **chemically combined**, each compound has different physical properties. On their own, some compounds are

- runny liquids;
- thick liquids;
- solids;
- gases.

Mixed up together they are a thick, gooey black liquid. As a mixture, crude oil is useless. The mixture must be separated into parts that are useful in an oil refinery.

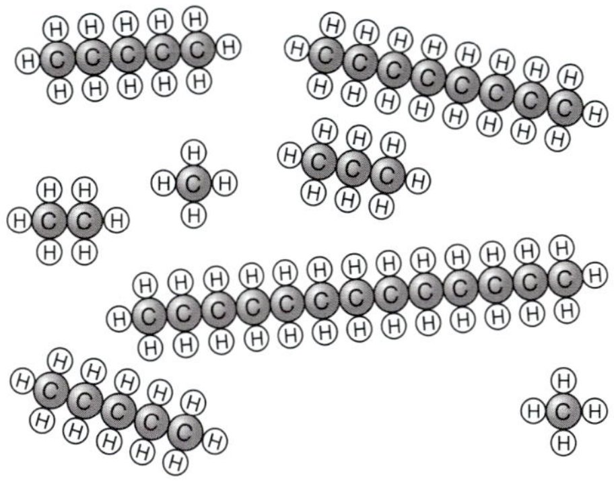

Crude oil is a mixture.

## What kind of compounds?

Most of the compounds in crude oil are made from two types of atom only: hydrogen and carbon. Compounds like these are called **hydrocarbons**.

These hydrocarbons have molecules of different sizes. Carbon atoms can form chains. Big molecules, with lots of carbon atoms, form long chains. Small molecules, with few carbon atoms, have short chains.

**d** Look at the crude oil mixture. How many carbon atoms are there in the chain of the longest hydrocarbon?

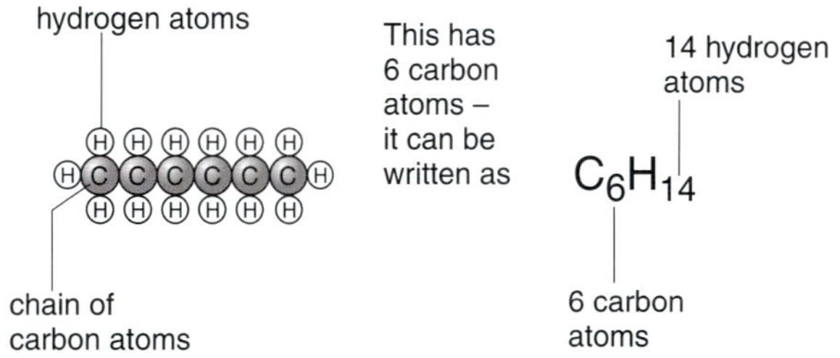


This hydrocarbon is called hexane.

## Questions

1 Copy and complete the following sentences. Choose words from this list.

**heated** **millions** **oxygen** **plants**

Crude oil formed from the remains of animals and ________ that lived ________ of years ago. They were buried in the rocks, where they were squashed and ________. Crude oil is formed when no ________ is there.

2 Crude oil is a ***mixture*** of different ***compounds*** of hydrogen and carbon. Explain the terms *mixture* and *compound*.

3 How many hydrogen atoms are there in a hydrocarbon molecule with three carbon atoms?

4 Which of these compounds are hydrocarbons?

$C_6H_6$ $C_2H_5OH$ $C_2H_2$ $CHCl_3$

### Key ideas

- *Crude oil is a fossil fuel.*
- *It formed from plants and animals that were buried in the rocks millions of years ago.*
- *Hydrocarbons are compounds of carbon and hydrogen only.*
- *Crude oil is a mixture of hydrocarbons.*

# Separating out

## Fractional distillation

Crude oil is a mixture of many useful hydrocarbons that need to be separated out before they can be used. This separation is carried out by a form of distillation.

If you heat crude oil enough it will boil. The gas can then be cooled so that it condenses back to a liquid for collection.

- Long-chain hydrocarbons with high boiling points will condense out quickly.
- Short-chain hydrocarbons with low boiling points will have to be cooled down much more before they condense.

**a Steam condenses back to water at 100 °C. Paraffin boils at 200 °C. At what temperature will paraffin gas condense back to liquid?**

As the gas cools, different **fractions** can be collected. The process is called **fractional distillation**.

## The fractionating tower

In an oil refinery, crude oil is heated strongly in a furnace so that it all boils. The gas then passes up through a fractionating tower, which is hot at the bottom but cold at the top.

The different hydrocarbons condense out at different levels of the tower and can be piped off.

The short-chain hydrocarbons remain as a gas as they float further up the tower. These will only condense when they get cool near the top of the tower.

**b Bitumen collects at the bottom of the tower. Does bitumen have long or short chains?**

This process can run continuously, which helps to keep the costs down.

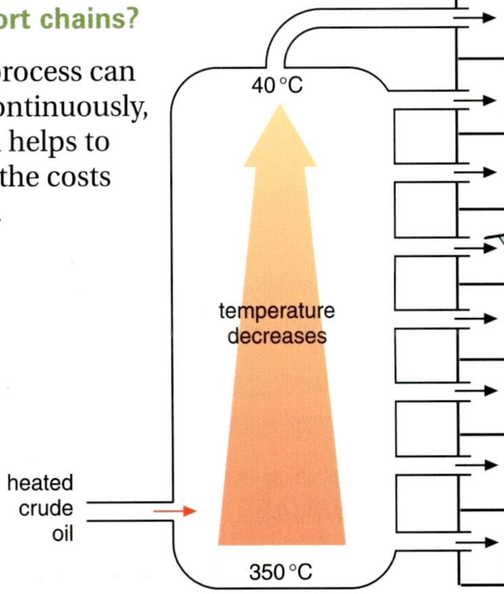


| | Fraction | Average number of carbon atoms |
|---|---|---|
| | **LPG (liquid petroleum gas)** is used for Calor gas stoves. | 3 |
| | **Petrol** is used as fuel for cars. | 8 |
| | **Naphtha** is used to make chemicals. | 10 |
| | **Paraffin** is used as a jet fuel as well | 12 |
| | **Diesel** is used as a fuel for lorries. | 20 |
| | **Fuel oil** is used for central heating. | 40 |
| | **Lubricating oil** is used to lubricate machinery. | 80 |
| | **Bitumen** is used to make roads. | 120 |

## The useful fractions

This process does not give a perfect separation, as each fraction is still a mixture. But the range of chain lengths within each fraction is small, so the hydrocarbons have very similar properties.

Petrol, for example, contains a range of molecules that have between 6 and 10 carbon atoms. In an engine the liquid petrol has to turn to a gas before it will burn. The hydrocarbons in this mixture turn easily to a gas.

**c** **In which fraction would you find a 75 carbon-chain hydrocarbon? What property makes this safer to store than petrol?**

**d** **LPG has to be stored under pressure in strong metal cylinders. Why would an LPG cylinder with a leaking valve be dangerous?**

### Did you know?

Heathrow airport near London uses huge amounts of kerosene as fuel for jet aircraft. There is an underground pipeline from the refinery near Southampton to supply this.

## Supply and demand

From 100 000 tonnes of crude oil you might get:

- 25 000 tonnes of petrol (25%);
- 15 000 tonnes of paraffin (15%);
- 25 000 tonnes of diesel (25%);
- 20 000 tonnes of fuel oil (20%).

**e** **Petrol makes up about 40% of all the crude oil products sold (by tonnage). Why might that cause problems for the oil refinery?**

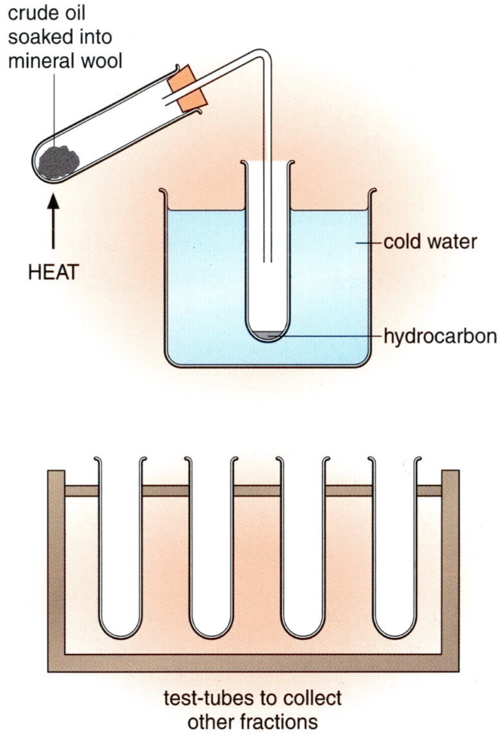


## Questions

**1** **Rearrange the sentences below to describe how fractional distillation works. Copy out this description.**

**A** **Long-chain hydrocarbons condense out at the bottom of the tower.**

**B** **The tower is hot at the bottom and cold at the top.**

**C** **Crude oil is turned into a gas in a furnace.**

**D** **Short-chain hydrocarbons condense out higher in the tower.**

**E** **The gas flows up through a fractionating tower.**

**2** **The diagram (above right) shows how you could get different fractions from crude oil in the school laboratory. You would need to change the collecting tube as the temperature rises. A different fraction is collected in each test-tube.**

**a** **Which word describes the method being used?**

**combustion** **distillation** **evaporation** **filtration**

**b** **Would you get long- or short-chain alkanes in the first sample collected?**

**c** **Why would it cost more to use this method on a larger scale than a fractionating tower for oil refining?**

### Key ideas

- *Crude oil is separated into useful products by fractional distillation.*
- *Each fraction has hydrocarbons of similar chain lengths and properties.*

# 4:3 Burning fuels

## Hydrocarbons as fuels

Most of the products of fractional distillation of crude oil are mixtures of hydrocarbons. All hydrocarbons burn.

### Dangers of hydrocarbons

Some hydrocarbons are very flammable. These are the ones with small short-chain molecules. It is dangerous to smoke at a petrol filling station because vapour from petrol is very flammable.

### Simple hydrocarbons

All of the names of the members of the alkane family end in –ane, e.g. methane. The prefix tells a chemist the number of carbon atoms.

**a** **Pentane is a hydrocarbon. How many carbon atoms do you think there are in it? (Hint: How many sides does a pentagon have?)**

## Complete combustion of hydrocarbons

When you burn a hydrocarbon what you end up with depends upon the conditions for burning. When you are using hydrocarbons as fuels, it is important to get the maximum amount of energy released. This is called **complete combustion**.

This happens when you burn the hydrocarbon in a plentiful supply of air or oxygen. The products are carbon dioxide and water. The word equation for the burning of methane is shown below.

methane + oxygen → carbon dioxide + water

**b** **Write a word equation for the complete combustion of ethane.**

## Incomplete combustion of hydrocarbons

You will get different products when you burn hydrocarbons in a limited supply of air or oxygen. The flame is often yellow in colour and soot (an impure form of carbon) is often formed.

**Incomplete combustion** of hydrocarbons can produce **carbon monoxide** and water instead of carbon dioxide and water. Carbon monoxide has the formula CO.

**c** **Write a word equation for the incomplete combustion of methane.**

Carbon monoxide is a very toxic gas. It is colourless and has no smell.

**d Why does the fact that carbon monoxide has no colour and no smell make it more dangerous?**

When a person breathes carbon monoxide into the lungs, it gets into the bloodstream. Haemoglobin in the blood transports oxygen around the body as oxyhaemoglobin. This breaks up easily when the oxygen is needed.

Carbon monoxide also reacts with haemoglobin, but it forms carboxyhaemoglobin. This does not split up, so blood is unable to transport oxygen.

The result of breathing in carbon monoxide is often death. The exhausts of cars are checked in annual MOT tests to ensure that CO emissions are not too high.

Waste gases leave the engine through the exhaust.

## Questions

1 Copy and complete these sentences. Use words from the list.

**complete dioxide incomplete monoxide oxygen**

When a hydrocarbon burns it uses up the gas ________ from the air.

If a hydrocarbon is burned in a lot of air, carbon ________ and water are produced. This is called ________ combustion.

If a hydrocarbon is burned with only a little air, carbon ________ and water are produced.

2 When a gas boiler is installed the engineer has to check the size of the air vent. Why is this important?

3 Copy and complete the word equations for the burning of octane.

A lot of air

Octane + ________ → ________ + ________

A shortage of air

Octane + ________ → ________ + ________

**Did you know?**

Carbon monoxide is dangerous in coal mines. Before carbon monoxide detectors had been developed, miners used to take canaries in cages down the coal mines with them. If carbon monoxide was present, the canary would fall off the perch and die. Being much smaller than the miners, the canary was affected first. This was the warning to the miners to get out quickly.

About 50 people die in the United Kingdom every year from carbon monoxide poisoning. This is often caused by faulty gas boilers and fires with inadequate ventilation. It can also occur if a car is run inside a closed garage.

## Key ideas

- *Hydrocarbons burn in a plentiful supply of air to form carbon dioxide and water.*
- *The maximum amount of energy is released when a hydrocarbon is burned in a plentiful supply of air.*
- *In a limited amount of air or oxygen, carbon monoxide and water are formed.*
- *Carbon monoxide is a toxic gas.*

# 4:4 Using fractions

## Liquid petroleum gases

Some gases come out of the top of the fractional distillation column without condensing. They can be used to provide the energy needed to run the refinery.

**a Where is a large amount of energy used in the refinery?**

These gases can also be sold as liquefied petroleum gas (LPG).

Cars with petrol engines can be converted to run on LPG. A large tank is put in the boot of the car and this can be filled with LPG at a few filling stations.

Part of a leaflet advertising conversion of cars to run on LPG is shown.

**b What is the advantage to a motorist of converting the car to use LPG?**

**c What are the disadvantages?**

Cars converted to use LPG produce fewer harmful gases in exhausts. This is of benefit to us all.

## Bitumen

Crude oil is a black gooey liquid. **Bitumen** is the remains of crude oil distillation when all the useful fractions have been removed.

Bitumen is very thick and almost solid. Nothing is wasted – it is not thrown away. It is used for resurfacing roads.

Often a layer of bitumen is spread on the road surface and covered with a thin layer of stone chippings. The bitumen binds the stones together.

**d What can happen to the bitumen on a hot day?**

## Cracking hydrocarbons

The fractional distillation of crude oil gives many useful products. Unfortunately the demand for these does not match the supply from the refinery.

The biggest demand is for short-chain alkanes like those in petrol, so oil companies have plenty of long-chain alkanes left over.

**e** There's not enough petrol from the refinery. What other product is there not enough of?

**f** For which fractions is there more coming from the refinery than can be sold?

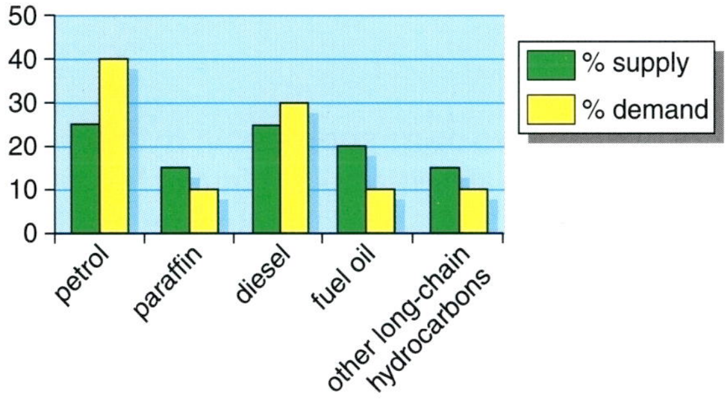


**Supply and demand for crude oil products.**

### Cracking

Fortunately, long-chain alkanes can be 'chopped up' into smaller, more useful pieces. If long-chain alkanes are heated, they can be broken up by a **thermal decomposition** reaction. This is called **cracking**.

**g** A 20 carbon-chain hydrocarbon was cracked to make an 8 carbon-chain hydrocarbon for petrol. How many carbon atoms would be left in the remaining molecule?

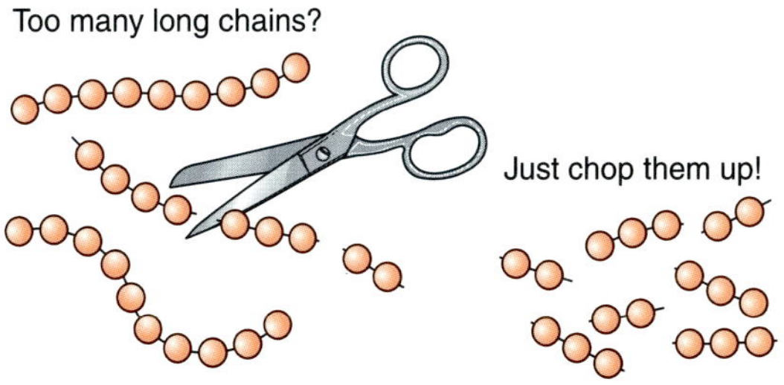


The small molecules made by cracking can be used to make plastics.

## Questions

**1** Finish the sentences by adding words from this list.

**cracking**    **high**    **hydrocarbons**    **low**    **plastics**

Long-chain ____________ have ______ boiling points. The demand for these materials is _______. However, splitting them up into smaller molecules by the process of __________ produces valuable materials that can be used to make _________.

**2** An oil company cannot make a profit selling petrol.

**a** Why is this statement true?

**b** Why does cracking help the oil company make a profit?

**3** Road surfacing is usually done in the summer months. Why is bitumen easier to use in summer?

**4** The names of three fractions from the fractional distillation column have become scrambled. Unscramble them and put them in order of increasing boiling point.

**SKREENOE**    **SIDEEL**    **MIEBNTU**

### Key ideas

- *Liquefied petroleum gas (LPG) can be used as an alternative fuel for cars.*
- *Bitumen is the remains from the fractional distillation column. It is not wasted. It is used for resurfacing roads.*
- *Long-chain molecules may be broken down into smaller molecules by cracking.*

# 4:5 Plastics

Cracking makes more short-chain hydrocarbons for fuels. It also makes small molecules that can be used to make other useful materials. The small molecules can be made to 'pop' together like beads to make very long chains called **polymers**. This is called **polymerisation**.

Polymers are also known as **plastics**.

Polymers are named by putting 'poly' in front of the name of the small molecule. The simplest molecule is called ethene, so the polymer made from this is called **poly(ethene)**. This is often shortened to polythene.

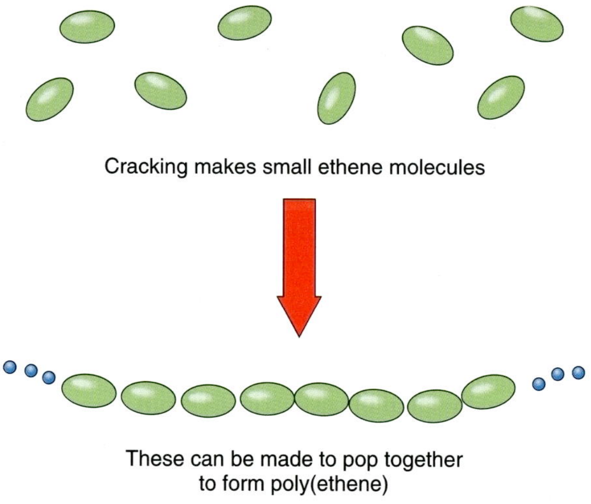


**a** **Another small molecule is called propene. What would the plastic made from this be called?**

**b** **PVC is a polymer. Its full name is poly(vinyl chloride). What is the name for the small molecules used to make PVC?**

## Poly(ethene)

Poly(ethene) is a waxy solid that is very easy to shape. It can be moulded into bottles for drinks or even dangerous chemicals such as bleach or acid. It can also be rolled into thin but tough, flexible and waterproof sheets. This makes it ideal as a food wrap or for plastic bags.

**c** **What advantage(s) do poly(ethene) bottles have over glass ones?**

## Poly(propene)

Poly(propene) is a tougher plastic than poly(ethene) and may be coloured easily too. It is used to make hardwearing things such as bowls, crates or even school chairs. It is also made into fibres for carpets and ropes.

**d** **Which polymer do you think washing up bowls are made from? Explain your answer.**

> **Did you know?**
> The polymer poly(ethene) was found by accident when chemists were trying to make other chemicals. The apparatus containing ethene was left and on return there was a waxy solid. This was poly(ethene).

## Uses of plastics

The table shows the properties and uses of some common polymers.

| Type of plastic | Properties | Uses |
|---|---|---|
| poly(ethene) PE | soft, flexible, waxy surface, translucent, melts at 80 °C, scratches easily | rubbish bags, squeeze bottles, dustbins |
| poly(propene) PP | hard but still flexible, waxy surface, melts at 145 °C, withstands solvents. Very versatile material with many applications | drinking straws, microwave ware, plastic garden furniture, baby baths, plastic lunch boxes |
| poly(vinyl chloride) PVC | flexible, clear, elastic, hardwearing | garden hose, shoe soles, cable insulation, watch straps |
| poly(styrene) PS | rigid, brittle, opaque, semi-tough, melts at 95 °C. Affected by fats and solvents | plastic cutlery, 'plastic glasses', low cost brittle toys, video cases, radiocasette player casings |
| expanded poly(styrene) EPS | foamed, light weight, energy absorbing, heat insulating | foamed polystyrene hot drink cups, foamed meat trays, protective packaging for fragile items |

**e** Which plastic can be used for packing some glass objects?

**f** Which plastic can be used for making the soles of shoes? What was used before plastics?

**g** Which plastic is used to make video cases and cheap toys? Why do they sometimes crack if they are dropped?

**h** Would you use poly(ethene) or poly(propene) to make a bowl for boiling water? Explain your answer.

## Questions

**1** Copy and complete these sentences. Use words from the list.

**molecules plastics polymers**

Some small __________ made by cracking can 'pop' together to make long chains called __________. Polymers are also known as ________.

**2** What is the name of the small molecule that is used to make poly(styrene)?

**3** Rearrange these sentences to explain how poly(ethene) is made from crude oil.

**A** The ethene is made to polymerise.

**B** Crude oil is fractionally distilled.

**C** Ethene is separated out.

**D** Some of the long-chain fractions are cracked.

### Key ideas

- *Some small molecules made during cracking are used to make plastics (polymers).*
- *Poly(ethene) is used for plastic bags and bottles.*
- *Poly(propene) is used for crates and ropes.*

# 4:6 Plastic problems

## Isn't it great!

Plastics are wonderful, aren't they? …

- bottles for our soft drinks;
- cartons for our hamburgers;
- cups for our coffee;
- plastic trays and cling film for meat;
- plastic foam for packaging;
- … and don't forget the plastic rubbish bags to throw it all away!

| Use | Percentage |
|---|---|
| packaging | 35 |
| building | 25 |
| electrical | 10 |
| transport | 7 |
| furniture | 7 |
| toys and leisure | 6 |
| other | 10 |

**How we use plastics**

**a Packaging is the major use of plastics. What was used for packaging before plastics?**

## The downside

Disposing of plastics is the problem. Over a million tonnes of plastic packaging are produced per year in the UK. A lot of this ends up as litter.

**b Suggest two ways in which plastic rubbish gets onto a beach.**

## Nature's way

Paper is thrown away too, but this natural material rots down rapidly in the soil – it is **biodegradable**. Plastics, however, are not biodegradable, and may take tens or even hundreds of years to break down.

**c People throw away banana skins and orange peel too. Why is this less of a problem than crisp packets?**

Most waste plastic does end up in the dustbin. It forms about 20% of the rubbish.

It then probably ends up on the local tip or landfill site.

But landfill sites are filling up fast. How long will it be before we drown in the sea of our own refuse?

**d Suggest two other ways in which we could dispose of our rubbish.**

## Different people, different solutions

**Earnest Green, environmental activist**
"We are being buried alive by plastic that pollutes our environment. It looks ugly and produces toxic chemicals when it breaks down. We must ban all plastic packaging now before it's too late. Some parts of New York State, USA, already have laws that have forced supermarkets to go back to using paper bags."

**Jayne Middleton, local campaigner**
"Plastics are produced from oil by processes that use large amounts of energy. What a waste to just throw it all away. The solution is to **recycle** wherever possible."

**Susan Foresight, chemist**
"Smarter technology can solve the problem. Chemical firms such as ICI are working on the development of new plastics which are biodegradable."

**Jack Burnham, combustion engineer**
"Plastic waste contains as much stored energy as fossil fuels. Why bother to recycle when you can just burn the plastic waste as fuel for a power station? Think how much coal and oil you could save."

## What do you think?

*Either:*

Draw up a table with two columns and list the advantages and disadvantages of plastics rather than 'natural' materials.

| Advantages | Disadvantages |
|---|---|
| | |

Design a poster to make people think about the problems caused by plastics. Use the points you have noted.

*Or:*

Working with a group, take on one of the characters quoted here. Try to develop their arguments by thinking the problems through and discussing them with your group.

You could try to find out more about the issues by visiting the library, searching the Internet or writing to environmental groups or industrial companies.

When you have gathered enough information, either write a report expanding on your ideas or hold a mock 'public enquiry' or debate on the way forward.

### Key ideas

- *Plastics can cause environmental problems because they are not biodegradable.*

# 4:7 Enzymes in brewing and baking

Enzymes are biological catalysts (Module 3 pages 88–89). All living cells use enzyme-controlled chemical reactions to make new materials.

Enzymes are widely used in many industries. Two industries that have used enzymes for centuries are brewing and baking.

**Did you know?**
There are over 30 references to wine in the Old and New Testaments of the Bible.

## Fermentation

Sugars are energy foods – body fuels. Your body reacts glucose with the oxygen you breathe to make carbon dioxide, water and the energy you need for life. Some organisms, such as **yeast** (a simple fungus), are able to use their special enzymes to get energy from sugars without oxygen, by the process of **fermentation**.

Many traditional festivals celebrate the importance of enzyme-based chemistry!

If you put living yeast cells into a sugary solution and keep it at about 37 °C, they start to grow and divide rapidly, turning the sugar into ethanol and producing lots of carbon dioxide which is allowed to escape.

$$\text{sugar} + \text{water} \xrightarrow[\text{yeast enzymes}]{\text{fermentation}} \text{ethanol} + \text{carbon dioxide}$$

This process has been used for centuries to make alcoholic drinks, such as wine and beer.

**a What are the conditions needed to ferment a mixture of grape juice and yeast to make wine?**

## Testing for carbon dioxide

Carbon dioxide is a colourless, odourless gas that puts out a burning splint.

If you want to be sure that a gas is carbon dioxide, you should use the *limewater* test. **Limewater** is a clear solution of calcium hydroxide. When you react it with carbon dioxide, you get insoluble calcium carbonate, which forms a white precipitate; this turns the solution 'milky'.

$$\text{calcium hydroxide} + \text{carbon dioxide} \rightarrow \text{calcium carbonate} + \text{water}$$

**b Describe a simple way in which you could test the gas given off by fermenting fruit juice to see if it was carbon dioxide.**

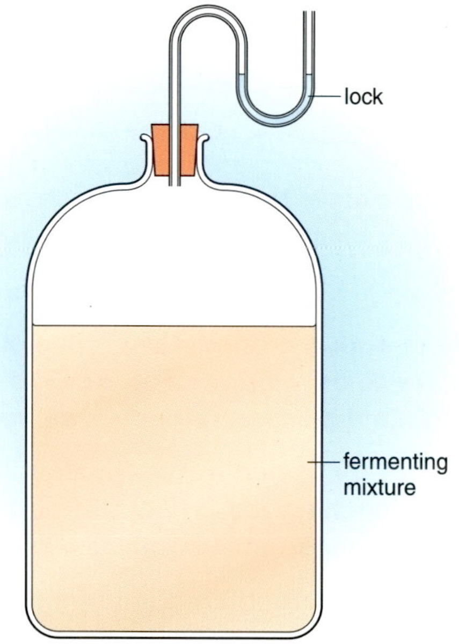


Wine fermenting. The lock allows carbon dioxide to escape but prevents air entering.

## Baking

The same enzyme-based reaction is also used in breadmaking. You mix yeast and a little sugar with water and flour to give an elastic dough. Then leave the dough in a warm place.

Fermentation begins but the carbon dioxide is trapped and cannot escape. It forms tiny bubbles within the dough, which get larger and larger, making the dough 'rise'. This makes light-textured bread when it is baked.

**c What effect will baking have on the enzymes as the temperature rises in the oven?**

The texture of the loaf comes from the fermentation of the dough producing bubbles of carbon dioxide.

### Raising agents

Some flour is called self-raising flour. You can use it, for example, when making cakes. This flour contains a chemical as a raising agent. This chemical is often sodium hydrogencarbonate, sometimes called bicarbonate of soda.

When this is heated in the oven it decomposes to form carbon dioxide and this forms bubbles within the cake as it bakes. This makes the cake rise.

$$\text{sodium hydrogencarbonate} \rightarrow \text{sodium carbonate} + \text{water} + \text{carbon dioxide}$$

**d Why is it unnecessary to leave a sponge cake to stand in a warm place before baking?**

**e Sam opens the oven door while a sponge is baking. The sponge does not rise. Suggest why.**

## Questions

1 Copy and complete the sentences. Use words from the list.

**carbon dioxide** **catalysts** **enzymes** **ethanol** **fermentation** **oxygen** **sugar**

Yeast contains biological _______ called _________ . When yeast is added to _________ solution, the process of ___________ can take place. This produces _________ and the gas _______________. This process takes place when the mixture is warm and _________ is absent.

2 Why is bread left to rise in a warm place for some time before baking? What would the best temperature be?

3 When bread rises it also produces ethanol. Why is there no ethanol in the bread after baking?

### Key ideas

- *Living cells use chemical reactions to produce new materials.*
- *Yeast cells convert sugar into carbon dioxide and ethanol.*
- *This process is called fermentation and is used:*
  - *to produce the ethanol in beer and wine;*
  - *to produce the bubbles of carbon dioxide that make bread dough rise.*
- *A simple laboratory test for carbon dioxide is that it turns limewater milky.*
- *Chemicals called raising agents are used to make cakes rise.*

# 4:8 More uses of enzymes

The use of enzyme technology has steadily increased over recent years as novel uses for enzymes have been discovered.

Bio yoghurts containing live bacteria.

## Cheese and yoghurt

Enzymes in yeast turn sugars into ethanol and release energy. Some bacteria get their energy in a similar way from the sugar in milk (lactose). The product is lactic acid rather than ethanol.

This makes the milk curdle to form cheese or yoghurt. You can buy some yoghurts with live bacteria in them.

If you want to make your own yoghurt, all you need to do is mix a little of this 'live' yoghurt into some milk and leave it in a warm place. The bacteria will grow rapidly and their enzymes will get to work on the milk. The milk provides the bacteria with food and energy and you with an endless supply of yoghurt.

**a What would be the best temperature to 'grow' your own yoghurt?**

**Did you know?**

People eat 'live' yoghurt to encourage special bacteria to grow in their stomachs. These 'good' bacteria help to stop other bacteria becoming established and causing food poisoning.

## Baby foods

The production of commercial baby foods is now very big business. At first, these were simply meals that had been pureed for small people without teeth. But very young babies can have difficulty digesting 'adult' foods, even in this form.

To overcome this problem, many baby foods are now 'pre-digested' using the enzyme **protease**. This starts the breakdown of the proteins in the food before it is put in the jar, so that young babies can digest it all more easily.

**b Protease breaks down proteins. What food group do you think is broken down by the enzyme carbohydrase?**

Baby foods are 'pre-digested' by protease.

## Biological washing powders

Many stubborn stains on clothing are caused by proteins, fats and oils that are not easily soluble in water. Traditional soap powders use soaps or detergents that help to loosen these chemicals and lift them free from the fabric.

A different approach is to use special enzymes that help to break up the chemicals before removing them. Protease is again used to break up proteins, while **lipase** is used to break up fats and oils. (Fats and oils are called *lipids*.)

**c** Some people do not like to use enzyme-rich washing powders in case they get them on their skin. Why might they be concerned about this?

**d** Biological washing powders give the best results at 40 °C rather than in boiling water. Suggest why.

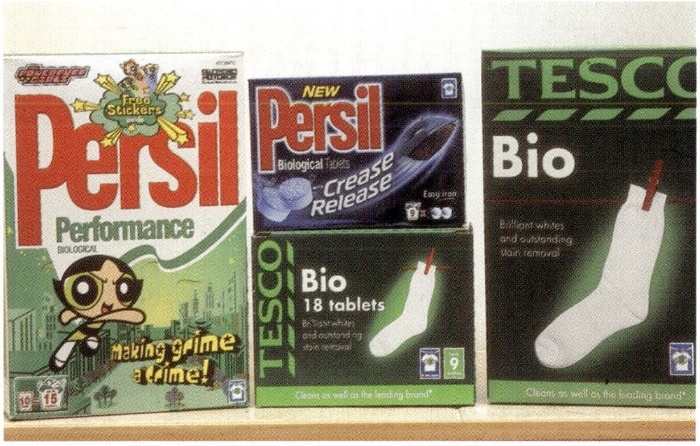


Washing powders may contain enzymes.

## Glucose

The sugar you have in your sugar bowl is called sucrose. This very sweet sugar is only extracted from sugar beet or sugar cane, which can make it relatively expensive.

The carbohydrate **starch** is much more common in nature and is produced in vast quantities in corn. Starch can be easily and inexpensively harvested and converted to the sugar **glucose** using the enzyme **carbohydrase**.

**e** Many sports drinks contain glucose. Why might athletes need extra glucose like this?

## Fructose syrup

The glucose syrup produced from cornstarch is not usually used directly in food. Glucose is not as sweet as sucrose so you would need to use more of it to get the same sweetening effect. This makes it very undesirable in slimming foods, for example.

**f** Why do people on diets want to reduce the amount of sugar they eat?

Fortunately another enzyme called **isomerase** can be used to convert glucose into another sugar called **fructose**. Fructose is even sweeter than sucrose, and so is used in smaller amounts.

## Questions

**1** Three types of enzyme are carbohydrase, lipase and protease.

Which type of enzyme is used to:

**a** remove an oil stain from a carpet;

**b** make glucose from carbohydrate in bread;

**c** remove bloodstains from a shirt?

**2** Soft-centred chocolates can be made by injecting hard-centred chocolates containing sucrose with invertase. This breaks the sucrose down into glucose and fructose, making the chocolate soft-centred. Copy and complete this word equation:

sucrose → __________ + __________

**3** Processes involving enzymes take place under warm conditions. Often alternative processes, without enzymes, take place at higher temperatures.

Suggest an advantage of enzyme processes.

### Key ideas

- *Bacteria produce yoghurt and cheese from milk.*
- *Enzymes are used:*
  - *to pre-digest baby food;*
  - *in biological washing powders to digest and remove stains;*
  - *to make glucose and fructose from cornstarch.*

# 4:9 In the limelight

You have probably heard the saying 'In the limelight'. It is used when somebody is in the news and people are talking about him or her. Where does this saying come from?

Before gas and electric lighting, theatres were lit by furnaces on stage with big blocks of limestone on them. The limestone gives out a dim light only. An actor on the stage was therefore 'in the limelight'.

A chemical reaction is taking place when limestone is heated. The limestone (calcium carbonate) is split up into calcium oxide and carbon dioxide. This is an example of **thermal decomposition**.

**Limestone quarrying.**

## Limestone

Limestone is a common rock with many uses. It usually formed in the sea from the remains of shells. Every year 150 million tonnes of limestone are quarried in Britain alone. This is used for buildings or road 'chippings', to make cement, concrete and glass and to neutralise excess acidity in soils (see Module 4.10).

**a Limestone is very useful – but would you like to live near a large quarry like this? List three possible problems.**

## Using limestone

Limestone has been used as a building stone for thousands of years.

Today, most limestone is used as a **raw material** to make something new. It is changed by chemical processes into other useful products such as cement.

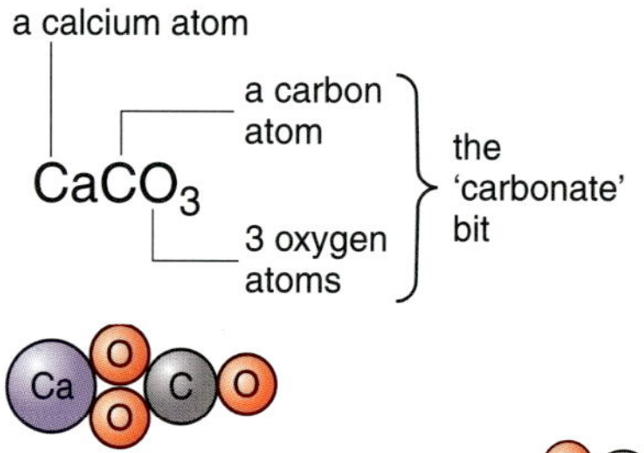


**A model of a calcium carbonate particle.**

## Processing limestone

Limestone is the compound **calcium carbonate**. Each particle has one calcium atom, one carbon atom and three oxygen atoms: $CaCO_3$.

If limestone is heated strongly thermal decomposition takes place. The compound breaks up and the atoms are rearranged.

- The carbon atom takes two oxygen atoms to form carbon dioxide. Carbon dioxide is a gas.
- The calcium atom is left with just one oxygen atom. This is **calcium oxide**. The common name for calcium oxide is **quicklime**.

$$\text{calcium carbonate} \xrightarrow{\text{heat}} \text{calcium oxide} + \text{carbon dioxide}$$

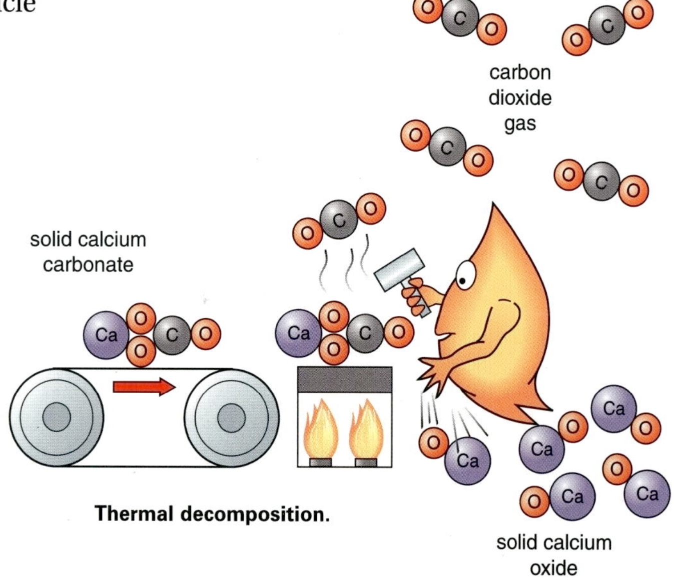


**Thermal decomposition.**

**b A manufacturer needs to buy limestone or quicklime for an industrial process. Suggest two reasons why quicklime will be more expensive to buy than limestone.**

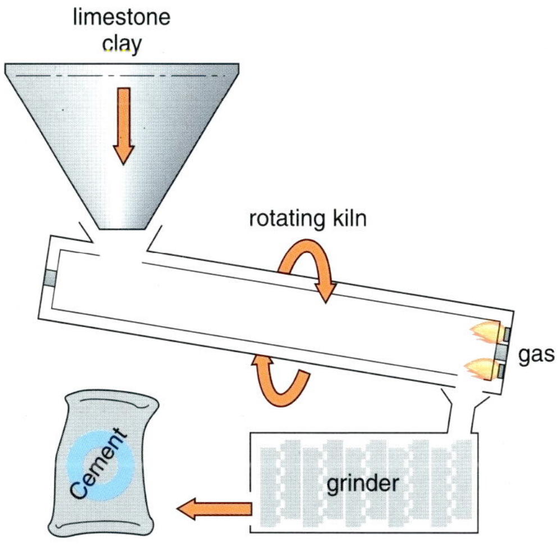


**A rotary kiln.**

## Cement and concrete

Today, most limestone is used to make **cement**. The limestone is heated with clay in a big oven that turns to mix them up. This type of oven is called a **rotary kiln**. This mixture is then ground to form a light grey powder. Cement turns back to a hard solid when you add water. This is a slow chemical reaction.

Cement can be mixed with sand and gravel to make **concrete**. This is cheaper and stronger than pure cement.

Concrete forms a thick liquid when first mixed with water, and can be poured into any shape. Slow chemical reactions make it set after a few hours and eventually it becomes rock-hard. It is used to make roads, bridges and the frameworks and foundations of buildings.

**c Why is it easier to use concrete for building than solid limestone?**

**Did you know?**
Coloured glass is formed when there are impurities in the glass.

## Glass

Pure sand is usually made from grains of **silicon dioxide**. **Glass** is made by heating very pure sand with limestone and a little soda (sodium carbonate). The chemicals react, melt and carbon dioxide is given off. When this liquid cools, it forms a hard but brittle, transparent material – glass!

calcium carbonate + silicon dioxide → calcium silicate + carbon dioxide

**d What is the chemical name for glass?**

## Thermal decomposition of copper carbonate

Copper carbonate is a green powder. When it is heated in a test-tube it splits up and a black powder is formed. Carbon dioxide gas is lost.

copper carbonate → copper oxide + carbon dioxide

**e If 2.0 g of copper carbonate is heated, will the mass remaining be more than 2.0 g, less than 2.0 g or exactly 2.0 g after heating?**

## Key ideas

- *When a substance is split up into two or more products by heating, it is said to have undergone thermal decomposition.*
- *Limestone and copper carbonate are split up by thermal decomposition into an oxide and carbon dioxide.*
- *Limestone is used as a building material.*
- *It is also converted into cement and concrete.*
- *Limestone is heated with sand to make glass.*

## Questions

1 Describe the ways in which limestone is used both in its natural state and as a raw material.

2 What raw materials are used in the manufacture of cement?

3 The photograph (above right) shows glass being collected for recycling. Why are different colours of glass collected in different bins?

# 4:10 Keeping in balance

## Controlling pH

Farmers growing crops on their land want to grow the highest yield they can. If they control the pH of their soil it can affect the amount they can grow.

Soil with a pH value of 6.6 to 7.0 is suitable for growing most plants. If the pH value goes below 6.0, the soil becomes too acidic for growing many plants.

If the pH value rises to 8.0 it will again be poor for plant growth. Very alkaline soil does not release vital minerals.

**a How could a farmer measure the pH value of the soil?**

In most cases soils become too acidic. This happens because alkalis are washed out of the soil by rain and rain is slightly acidic. Atmospheric pollution also makes the soil more acidic.

The removing of the excess acid from the soil to restore a balance between acidity and alkalinity is called **neutralisation.**

**Did you know?**
If we were able to control the pH values of soil it is estimated that the world could grow 20% more food.

## Reducing acidity

A farmer wanting to reduce the acidity of the soil is likely to lime the soil in autumn. Calcium oxide could be used but it is difficult to store and handle.

Instead farmers use calcium hydroxide. The common name for calcium hydroxide is slaked lime. This is formed when quicklime (see page 110) is added to water.

calcium oxide + water → calcium hydroxide

**b Suggest why calcium hydroxide is quick acting.**

In addition to reducing acidity, calcium hydroxide improves the qualities of the soil and provides calcium required for plant growth.

Calcium carbonate (crushed limestone) is an alternative way of controlling excess acidity. Because it is much less soluble than calcium hydroxide, it remains in the soil for much longer.

## Using fertilisers

In addition to controlling the pH value of the soil, the farmer should add essential elements needed to get plants to grow well. This can be done with natural animal wastes such as manure. Alternatively, chemical **fertilisers** can be added.

**c A fertiliser bag has the letters NPK on it. Which chemical elements are in this fertiliser?**

Ammonium nitrate, ammonium sulphate and potassium nitrate are all fertilisers. They are also **salts**.

## What are salts?

A salt is the product formed when an acid reacts with a base or alkali. Common salts are chlorides, sulphates, nitrates, carbonates and phosphates.

This table shows the acids required to make different salts.

| Salt | Acid used |
|---|---|
| chlorides | hydrochloric acid, $HCl$ |
| sulphates | sulphuric acid, $H_2SO_4$ |
| nitrates | nitric acid, $HNO_3$ |
| carbonates | carbonic acid, $H_2CO_3$ |
| phosphates | phosphoric acid, $H_3PO_4$ |

Ammonium nitrate is manufactured from nitric acid and ammonia solution (an alkali).

ammonia solution + nitric acid → ammonium nitrate + water

Ammonium nitrate is made into tiny solid granules.

**d What is used to manufacture ammonium sulphate?**

The general equation for salt formation is:

acid + base (or alkali) → salt + water

**Did you know?**
A solution of calcium hydroxide in water is called limewater. Calcium hydroxide is quick acting.

## Questions

1 Copy and complete these sentences. Use words from the list.

acidic alkaline calcium hydroxide
fertiliser nitric acid nitrogen

Farmers may add lime (__________) to soil to make it less ________.

Ammonium nitrate is a __________ used to supply __________ to plants.

Ammonium nitrate is formed when ammonia solution reacts with ______________.

2 a What reacts with sodium hydroxide to form sodium nitrate?

b Write a word equation for the reaction.

3 Ammonium phosphate gives two important elements to the soil. Which two elements?

### Key ideas

- *Acidic soils are neutralised using an alkali such as slaked lime or calcium carbonate.*
- *Plants grow best in soils that are approximately neutral.*
- *Fertilisers are used to add to soils the elements that are needed by plants to grow.*
- *The three elements needed in large quantities by plants are nitrogen, phosphorus and potassium.*
- *Fertilisers are usually salts.*

# End of module questions

## Low demand

1. Crude oil is a mixture of:

   **A** carbon atoms

   **B** elements

   **C** gases

   **D** hydrocarbons.

2. The drawings show four different molecules.

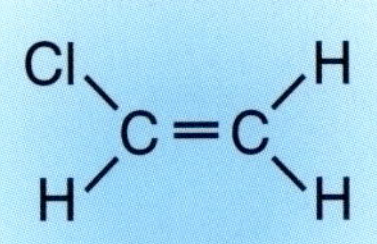

**W**

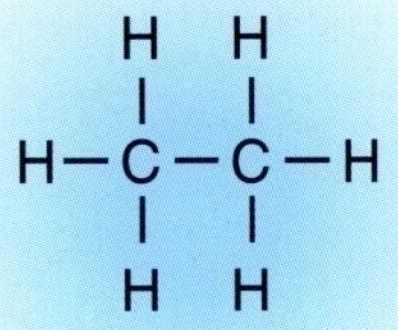

**X**

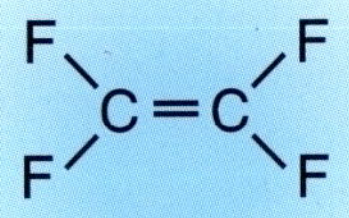

**Y**

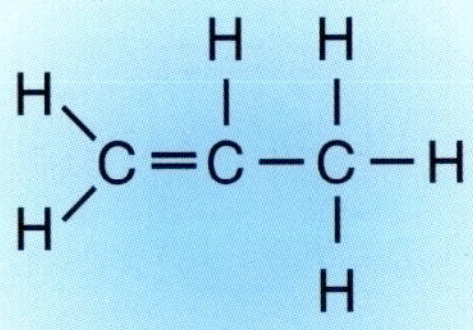

**Z**

   Which two molecules are hydrocarbons?

   **A** W and X

   **B** X and Z

   **C** X and Y

   **D** Y and Z

3. Which of the following are made from limestone?

   **A** cement

   **B** copper metal

   **C** detergents

   **D** plastics

4. Poly(propene) is used to make:

   **A** bicycle frames

   **B** bridges

   **C** overhead power cables

   **D** washing up bowls.

5. Fertilisers such as ammonium sulphate are made by:

   **A** combustion

   **B** decomposition

   **C** distillation

   **D** neutralisation.

6. Biological washing powders contain materials to remove stains from coloured clothes.

   These materials are:

   **A** acids

   **B** bleaches

   **C** enzymes

   **D** indicators.

7. Ethane is a hydrocarbon. Which is the correct molecular formula for ethane?

   **A** $CH_4$

   **B** $C_2H_6$

   **C** $C_3H_8$

   **D** $C_4H_{10}$

8. Plastics cause problems when we want to get rid of them because they:

   **A** do not burn

   **B** do not rot away

   **C** are lightweight

   **D** are made from crude oil.

9. Enzymes are:

   **A** acids

   **B** alkalis

   **C** biological catalysts

   **D** salts.

**10** Crude oil was made in the Earth from:

**A** coal

**B** dead sea creatures

**C** rocks

**D** salt.

**11** Crude oil is separated into useful products by:

**A** combustion

**B** evaporation

**C** filtration

**D** fractional distillation.

**12** The process in question 11 is carried out in a:

**A** brewery

**B** laundry

**C** refinery

**D** refractory.

## Standard demand

**13** The reaction that produces calcium oxide from limestone is an example of:

**A** combustion

**B** cracking

**C** fractional distillation

**D** thermal decomposition.

**14** Which of the following word equations shows the reaction that produces slaked lime?

**A** calcium carbonate + water → calcium hydroxide

**B** calcium hydroxide + water → calcium oxide

**C** calcium hydroxide → calcium oxide + water

**D** calcium oxide + water → calcium hydroxide

**15** Hydrocarbons with small numbers of carbon atoms:

**A** are produced at the bottom of the column

**B** are thick and difficult to pour

**C** catch light easily

**D** have high boiling points.

**16** Which of the following is produced when hydrocarbons burn in a plentiful supply of air?

**A** carbon

**B** carbon dioxide

**C** carbon monoxide

**D** hydrogen

**17** When calcium carbonate is heated a gas is given off. The gas is:

**A** carbon monoxide

**B** carbon dioxide

**C** oxygen

**D** ozone.

**18** Cement is made by:

**A** heating limestone and adding water

**B** heating limestone with clay

**C** dissolving limestone in acid

**D** mixing limestone with slaked lime.

**19** Some people think plastic polymers are environmental hazard because they are:

**A** made from crude oil

**B** not biodegradable

**C** easily moulded

**D** poor conductors of electricity.

**20** Oxidation is a reaction where:

**A** hydrogen is added

**B** oxygen is added

**C** oxygen is removed

**D** water is removed.

**21** Which row of the table is correct?

| | Where is bitumen obtained from the column? | Use of fraction |
|---|---|---|
| A | top | car fuel |
| B | middle | making rubber tyres |
| C | bottom | aircraft fuel |
| D | bottom | road surfacing |

22 The table gives the boiling points of four hydrocarbons. Which hydrocarbon is made up of the smallest molecules?

| Hydrocarbon | Boiling point in °C |
|---|---|
| A | –89 |
| B | –1 |
| C | –162 |
| D | –42 |

23 Which substance may be formed when methane burns in a limited amount of air but not when it burns in a plentiful supply of air?

**A** carbon dioxide

**B** carbon monoxide

**C** hydrogen

**D** water

24 How many atoms are bonded together in a molecule of ethane?

**A** 2

**B** 4

**C** 6

**D** 8

# Module 5 – Energy and electricity

Electricity is essential to the way we live our everyday lives. It provides:

- heat – for cooking and keeping warm;
- light – so that we are not restricted to being active only during the daylight hours;
- movement – for devices such as washing machines and computers that carry out everyday tasks for us.

In this module you will learn about different components used in circuits and how devices transfer energy from the mains electricity supply.

Mains electricity can be dangerous so it is important that you learn how to use it safely. You also need to be aware of where the energy for generating electricity comes from and how new sources of energy will have to be developed during your lifetime.

Try these questions first.

1 What does an ammeter measure?

2 How should an ammeter be connected in a circuit, in series or in parallel?

3 An electric circuit consists of a battery, a lamp and a motor. Which of these is the energy source?

4 Which of these energy sources is renewable?
Choose one from the list.

biomass coal gas oil

5 Explain whether wood is a renewable or a non-renewable source of energy.

# 5:1 Circuits

We use electric circuits every day. Some circuits, like that in a torch, have just a few components. The circuit diagram shows how they are connected together.

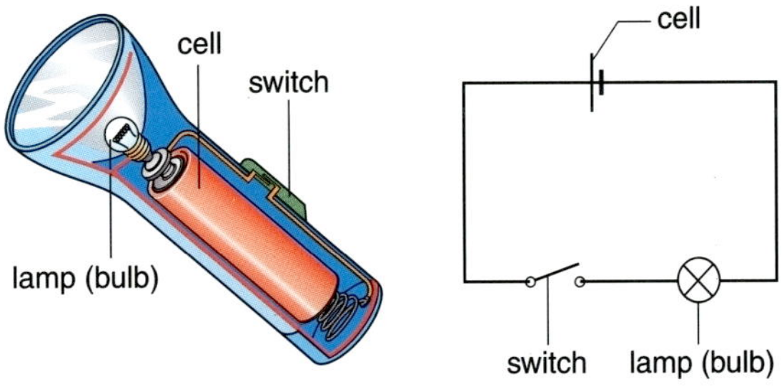


**This torch has just three components, connected in series.**

**A computer is a complicated circuit, with many different types of component.**

Other circuits, like those in a computer, have hundreds or thousands of components.

**a Draw and label the circuit symbols for a cell, a filament lamp and a switch.**

## Driving a current

In a torch, the components are connected in **series** (end-to-end). The cell provides the **voltage** needed to cause a **current**. Torches and personal stereos usually have more than one cell. Two or more cells connected together make a **battery**.

In the batteries used in torches, a single cell provides a voltage of 1.5 volts (1.5 V).

When cells are connected correctly in series, their voltages add together.

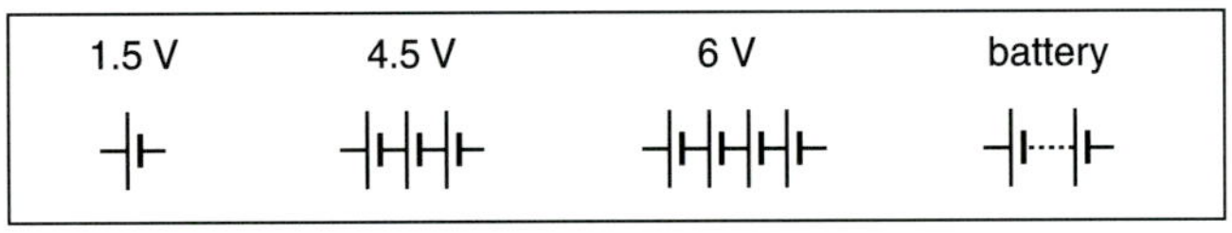


**Cells are joined together to make a battery.**

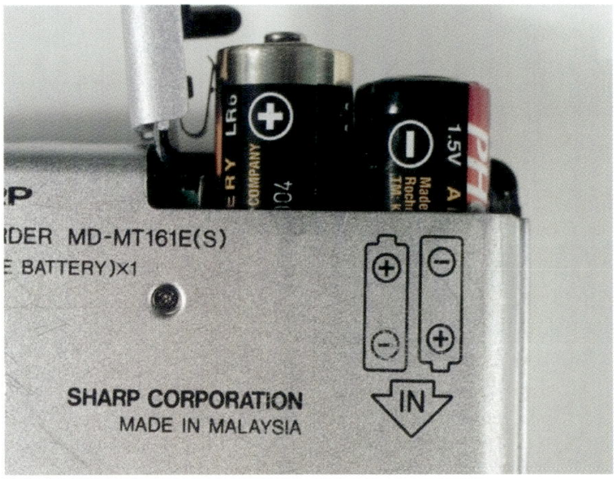


**The diagram shows how to put the cells the right way round in this personal stereo.**

**b How many cells need to be connected in series to make a 9 V battery?**

> **Did you know?**
> The cells in a car battery each have a voltage of 2 V.
>
> Six cells are connected in series to make a battery.

## Measuring current

Electric current is measured in amps (A) using an **ammeter**. An ammeter is always connected in series with the current being measured. The circuit diagram shows how to connect two ammeters in a circuit to measure the current passing into and out of a lamp.

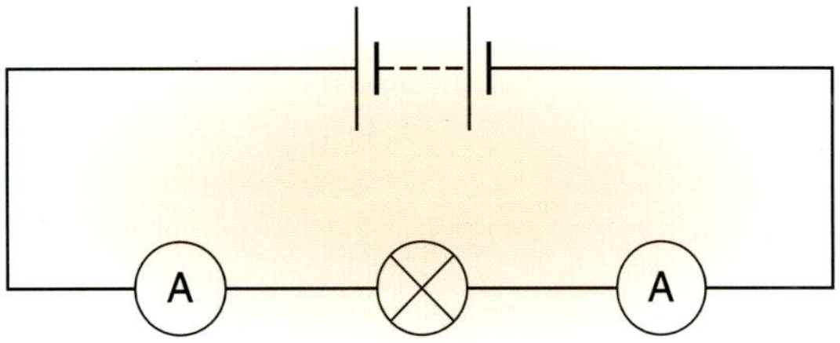


The size of the current that passes in a circuit depends on the **resistance** and the voltage. If the resistance is low, there is little opposition to the current. A circuit that has a high resistance makes it much harder for the current to pass.

- Increasing the resistance of a circuit makes the current smaller.
- Increasing the voltage makes the current larger.

## Measuring voltage

Voltage is measured using a **voltmeter**. This is always connected in **parallel** with a battery, power supply or circuit component. The diagram shows a circuit with two voltmeters. One is measuring the battery voltage and the other is measuring the voltage across one of the lamps.

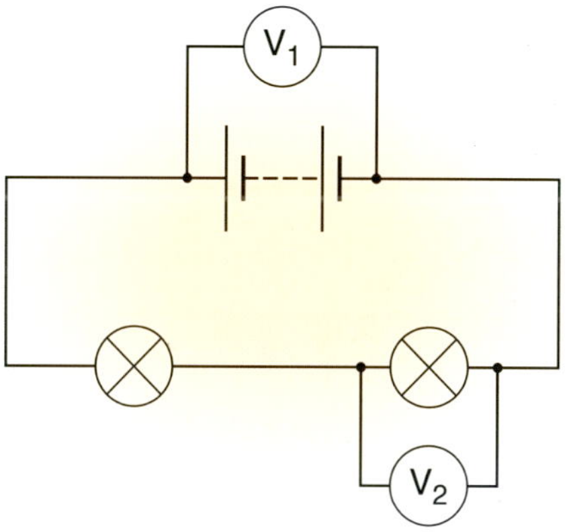


**c Which voltmeter is measuring the battery voltage and which is measuring the voltage across a lamp?**

## Questions

**1** Look at the diagram of a circuit.

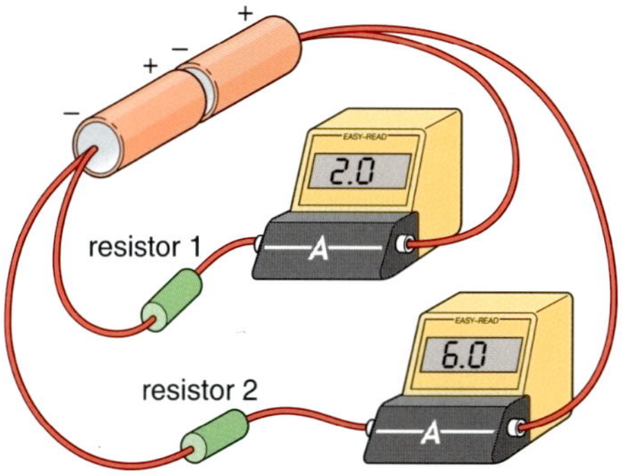


**a** Are the resistors connected in series or in parallel?

**b** Which resistor has the greater resistance?

Explain how you can tell.

**2** This is the circuit symbol for a resistor.

**a** Draw a circuit diagram that shows two resistors connected in parallel to a battery.

**b** Add voltmeters to your diagram that measure the voltage across:

**i** the battery

**ii** each of the resistors.

## Key ideas

- *The current in a circuit is measured in amps, using an ammeter placed in series.*
- *The voltage across a battery or component is measured in volts, using a voltmeter connected in parallel.*
- *The current that passes in a circuit depends on both the voltage and the resistance.*

# 5:2 Current, voltage and resistance

Everything has resistance. In a circuit the resistance of the voltage supply and the connecting wires is usually low. This means that we need only consider the resistance of the components that make up a circuit.

## Changing the resistance

Components can be added to a circuit in different ways.

When components are added in series:

- the resistance of the circuit increases
- less current passes in the circuit.

When components are added in parallel:

- the resistance of the circuit decreases
- more current passes in the circuit.

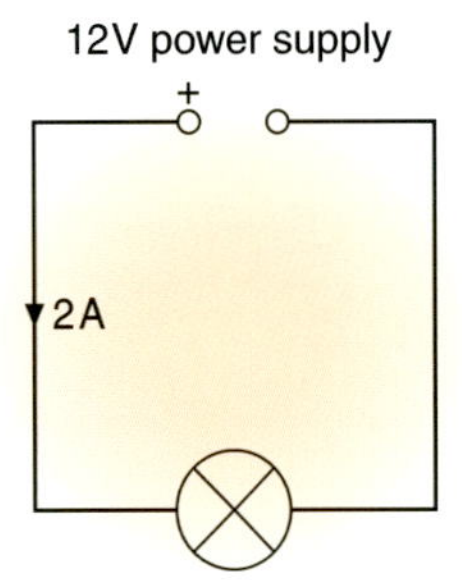

A current of 2 A passes in the lamp.

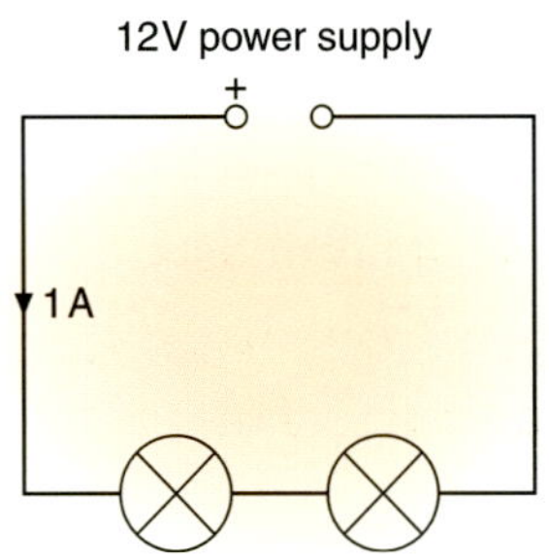

With two lamps in series, the current is only 1 A.

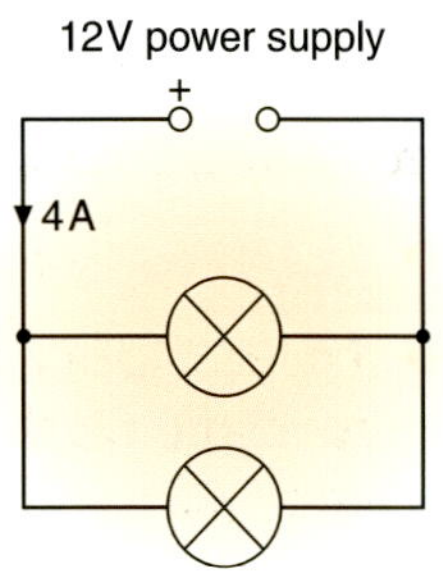

When the lamps are connected in parallel, the current is 4 A.

A **variable resistor** can also be used to change the current in a circuit. It has a slider that is used to change its resistance.

- A variable resistor connected in series in a circuit can be used to increase or decrease the current.

The circuit diagram shows how a variable resistor is used in a lamp-dimming circuit.

The diagram shows a variable resistor and the slider that is used to change the resistance.

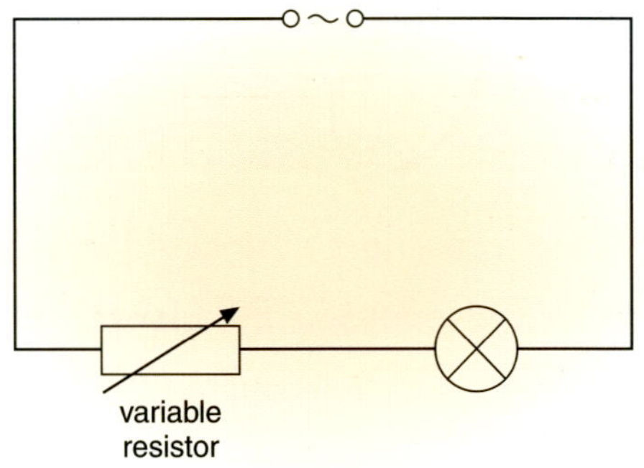

**The variable resistor is used to make the lamp brighter or dimmer.**

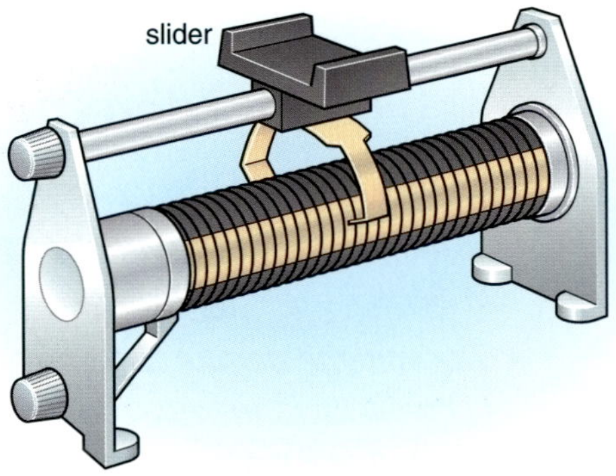

**A variable resistor.**

> **Did you know?**
> Variable resistors are used as the volume controls on televisions and hi-fi systems.

**a Explain how the brightness of the lamp changes when the resistance of the variable resistor is increased.**

## Using resistors

People who design electric circuits use resistors in the circuits.

Resistors are used to:

- control the current that passes in a component;
- fix the voltage across a component.

The circuit designer needs to know how much current passes in a resistor. The picture shows a designer using a computer to work this out.

**b How can the current passing in a component be made smaller?**

## The variables

There are three variables in a circuit.

- The voltage of the power supply or battery.
- The resistance of the circuit. This depends on the resistance of the components and how they are connected.
- The current that passes in the circuit. This depends on both the voltage and the resistance.

Voltage is measured in volts (V). Resistance is measured in ohms (Ω) and current is measured in amps (A).

These three variables are related by this formula:

Voltage = current × resistance  or  $V = I \times R$

**Example:** The lamp in the picture has a resistance of 4 Ω.

What voltage is needed to cause a current of 3 A to pass in it?

Step 1: Write down the formula, in words or symbols.

Voltage = current × resistance

Step 2: Write out the formula with the known values in place of the words.

Voltage = 3 A × 4 Ω

Step 3: Work out the answer and state the correct unit. In this case the answer is the Voltage, so the unit is volts.

Voltage = 12 V.

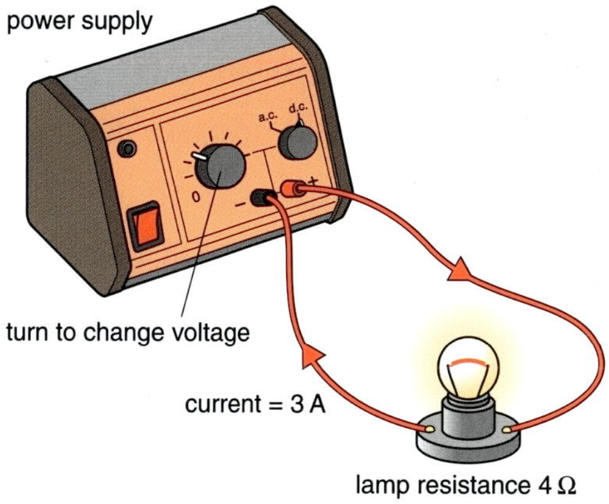


## Questions

**1** The current in a kettle element is 10 A. The element has a resistance of 24 Ω. Calculate the voltage across the element.

**2** A 6 V lamp and a 12 V lamp have the same current passing in them when connected to their normal operating voltages. Which lamp has the greater resistance? Explain how you can tell.

### Key ideas

- *A variable resistor can be used in series to change the current in a circuit.*
- *The relationship between current, voltage and resistance is:*

  *Voltage = current × resistance.*

# 5:3 Resistance and wires

The sound technician moves the sliders to adjust the sound coming from each loudspeaker. Moving the slider up increases the voltage, so more current passes in the loudspeaker and the sound is louder.

You can change the brightness of a lamp by changing the voltage of the power supply. When the voltage is increased, the lamp gets brighter. The current in the lamp has increased.

**a What happens to the brightness of the lamp when the voltage is decreased? What happens to the current?**

## Increasing the voltage

You can use this circuit to find out how the current in a **resistor** depends on the voltage across it.

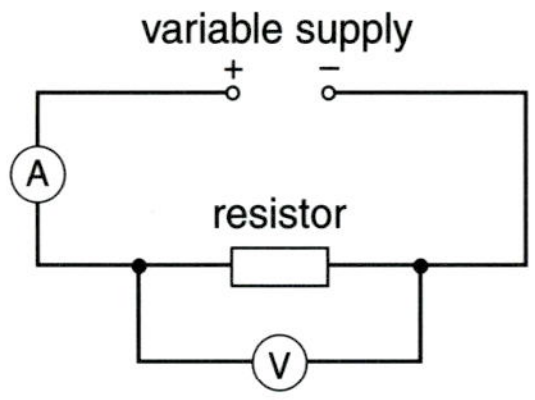


- The power supply is adjusted to vary the voltage.
- The meters show the voltage across the resistor and the current passing in it.

Some results of this experiment are shown on the graph.

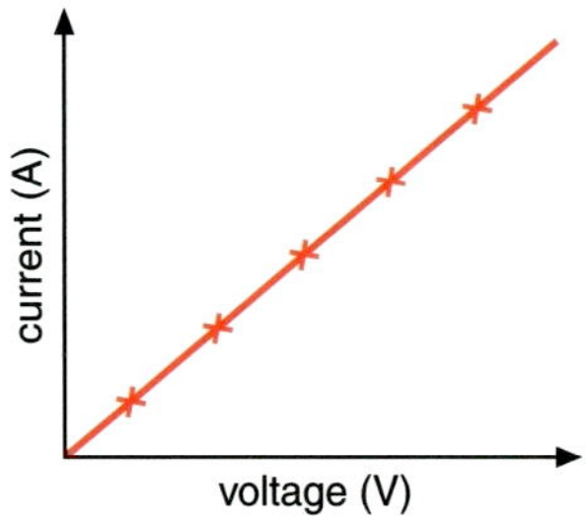


The graph is a straight line. It shows that:

- The current increases steadily as the voltage increases.
- When the voltage is doubled, the current also doubles. The current and voltage are said to be in **proportion**.

## Current in reverse

Turn off the power supply, swap over the positive and negative connections and turn the power back on again. This makes the current pass in the opposite direction. The voltage and the current are now negative. If you measure the current at the same voltage settings as before, the readings are similar.

Here is a completed graph, showing the current for both positive and negative voltages.

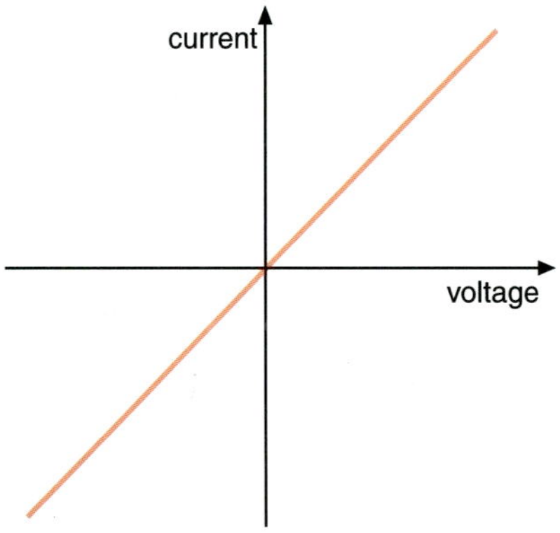


**Resistor (at constant temperature).**

**The graph of current against voltage for a fixed value resistor is a straight line.**

The graph shows that the same voltage applied in either direction causes a current of the same value. Reversing the voltage just reverses the direction of the current.

**b Copy and complete this sentence to describe the graph.**

**When the voltage across the resistor is increased in equal steps, the current in it also increases __________ .**

## A filament lamp

A lamp contains a thin wire called a **filament**. When it is switched on, the current that passes in the filament causes it to get hot enough to give out light. The greater the current in the filament, the hotter the filament becomes. As a wire heats up, its resistance increases.

The graph of current against voltage for a filament lamp is a curve.

- As the voltage across the lamp is increased in equal steps, the current increases in smaller and smaller steps.
- As the voltage increases the slope of the curve decreases.
- These effects are due to the increasing resistance of the filament.

**c What does the graph show about the behaviour of the lamp filament when the voltage is reversed?**

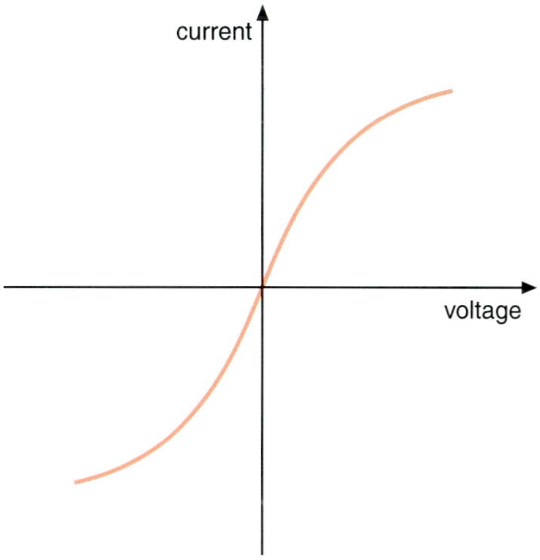


**The graph shows current against voltage for a filament lamp.**

**Did you know?**
The first filament lamp was invented by Thomas Edison in 1879. The filament was made of carbon. Since 1907 tungsten has been used for lamp filaments.

## Questions

1 The circuit in the diagram is used to see how the current in component X changes when the voltage is changed. The graph shows the results.

Copy the following sentences, choosing the correct word from each pair.

a Meter $M_1$ is a voltmeter/ammeter.

b Meter $M_2$ is a voltmeter/ammeter.

c The component X is a lamp/resistor.

d The graph shows that the current in X increases/decreases as the voltage across it increases.

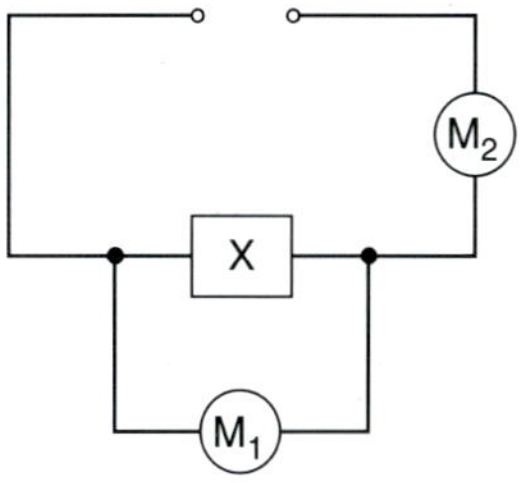


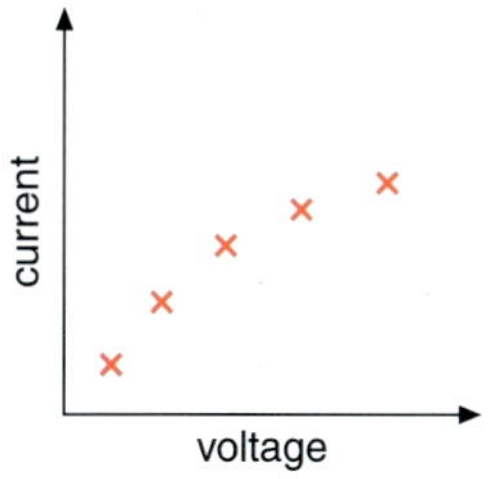


2 The table shows how the current in a resistor changes when the voltage across it is changed.

a Draw a line graph of current against voltage.

b Read from your graph the value of the current in the resistor when the voltage across it is:

i 2.5 V

ii 5.0 V

c What pattern do the answers to part b show?

| Voltage in V | Current in A |
|---|---|
| 0.0 | 0.0 |
| 2.0 | 1.6 |
| 4.0 | 3.2 |
| 6.0 | 4.8 |
| 8.0 | 6.4 |
| 10.0 | 8.0 |

### Key ideas

- *The current in a resistor is proportional to the voltage across it.*
- *The resistance of a filament lamp increases as the current in the filament increases.*

# 5:4 Three useful devices

A security engineer fits safety devices to houses, schools, offices and factories. How do these work?

- A fire alarm has a sensor that detects a change in temperature.
- A security light has one sensor that senses whether it is light or dark and another sensor that detects the presence of a person.

## Feeling the heat

A **thermistor** is a type of variable resistor. Its resistance changes when its temperature changes. This makes it useful as a temperature sensor. The graph shows how the resistance of a thermistor depends on its temperature.

This graph shows that:

- when the thermistor is cold, its resistance is high;
- as the temperature of the thermistor rises, its resistance decreases.

A thermistor.

The circuit symbol for a thermistor.

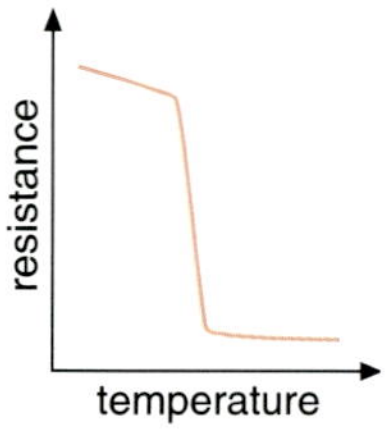


A fire alarm uses a thermistor as a temperature sensor. As the temperature rises, the current in the thermistor increases and this triggers the alarm.

**a Copy and complete this sentence.**

**As the temperature increases, the resistance of a thermistor __________ , so the current passing in the thermistor __________ .**

Thermistors are very sensitive. They can detect the heat from a candle flame several miles away.

## Seeing the light

A **light-dependent resistor** (LDR) is another type of variable resistor. It can be used to detect whether it is light or dark. The graph shows how the resistance of an LDR depends on the light level.

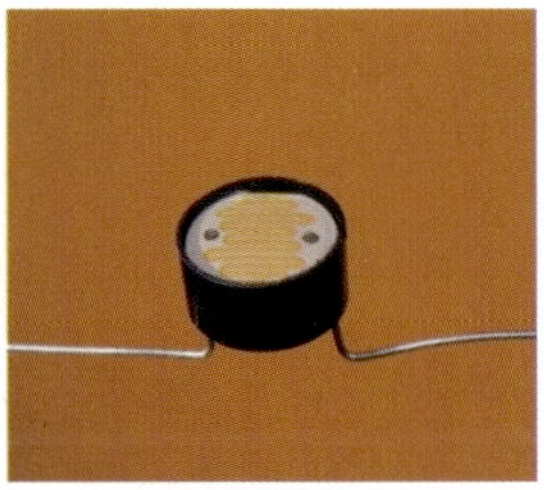

A light-dependent resistor.

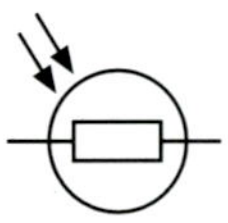

The circuit symbol for a light-dependent resistor.

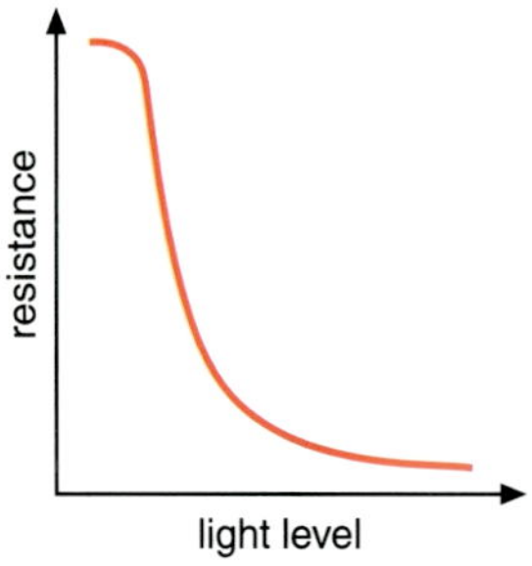


The graph shows that:

- when the LDR is in the dark, its resistance is high;
- when bright light shines on the LDR, its resistance is much lower.

This security light uses an LDR. At night, less light falls on the LDR. Its resistance increases. This switches on the light.

**b** Copy and complete the sentence.

When more light falls on an LDR, its resistance ________ so the current passing in it ________ .

## One way traffic

A **diode** is a device that allows current to pass one way, but not the other. Diodes are used to change alternating current from a transformer into the direct current needed by devices such as keyboards and computers. If a diode is connected the wrong way round in a circuit, no current can pass.

The arrow on the circuit symbol shows which way current can pass in a diode.

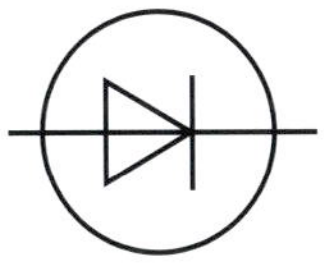

**This is the symbol for a diode. The arrow shows the direction in which current is allowed to pass.**

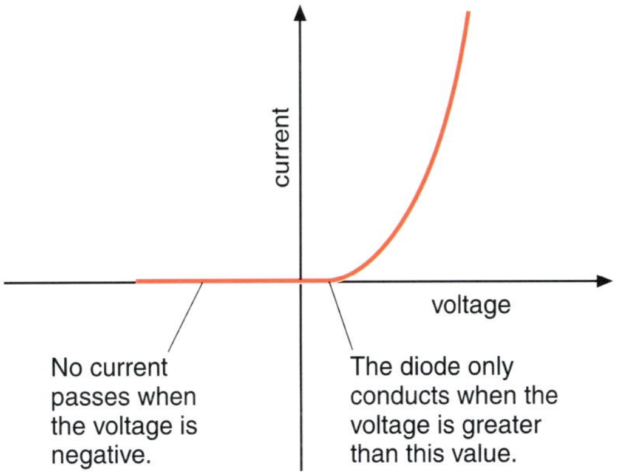


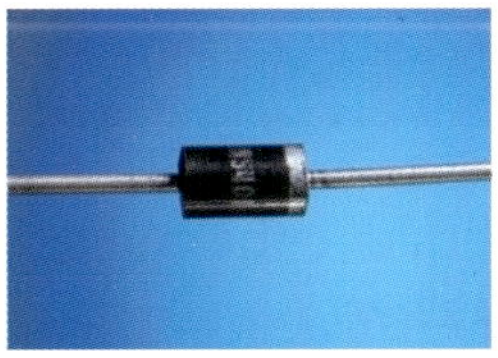

**A diode.**

The graph of current against voltage for a diode shows that:

- a minimum voltage is needed for the diode to conduct in the forwards direction;
- as the voltage increases above the minimum, the resistance of the diode becomes less;
- the diode does not conduct in the reverse direction.

**c** How can you tell from the graph that a diode has a very high resistance in the reverse direction?

## Questions

**1** Copy the circuit symbols and label each one with the correct name from the list:

**diode** **thermistor** **light-dependent resistor (LDR)**

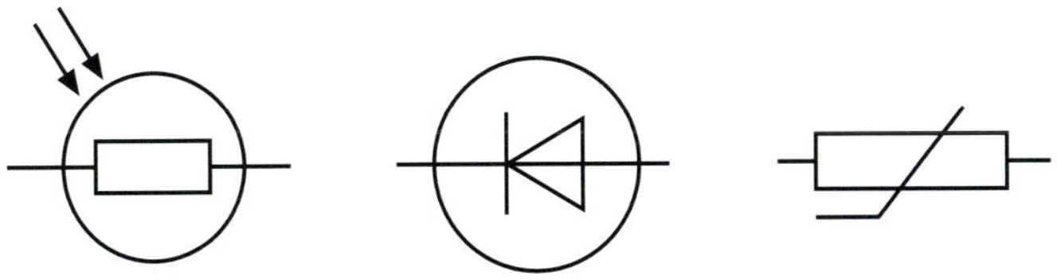

**2** Write out these sentences, choosing the correct word from each pair.

**a** The resistance of a thermistor increases/decreases when it is heated.

**b** The resistance of an LDR increases/decreases when light falls on it.

**c** A diode only allows current to pass in the forward/reverse direction.

## Key ideas

- *The resistance of a thermistor decreases as its temperature increases.*
- *The resistance of a light-dependent resistor (LDR) decreases as the light intensity increases.*
- *A diode allows current to pass in one direction only.*

# 5:5 Wiring up safely

The mains electricity supply can be dangerous. In the UK, its voltage is 240 V. If you touch the live wire, the current that passes in your body will give you a nasty shock. If the current is large enough it can cause serious injury, and it may be enough to kill you.

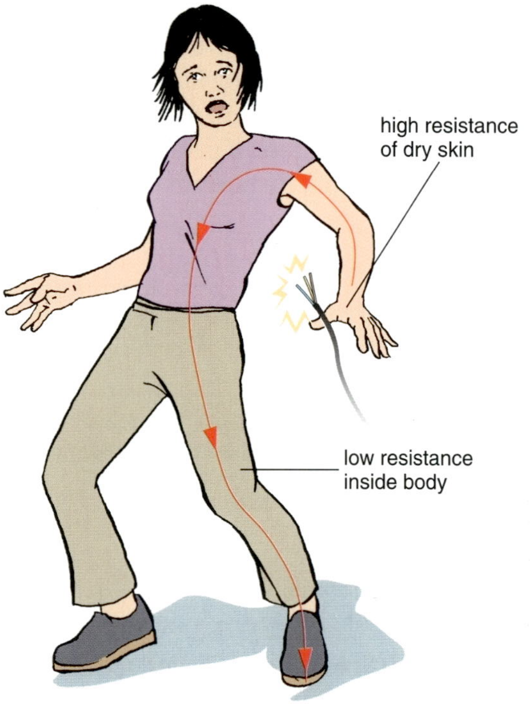


**Dry skin has greater resistance than wet skin. This may be enough to save you from a fatal shock.**

## Plugs and cables

Today, most electrical appliances are sold with a **plug** already fitted to the end of the cable. This helps to ensure the user's safety.

Features of the cable are:

- the metal wires are made of copper (a good **conductor**);
- the wires are surrounded by flexible plastic (a good **insulator**).

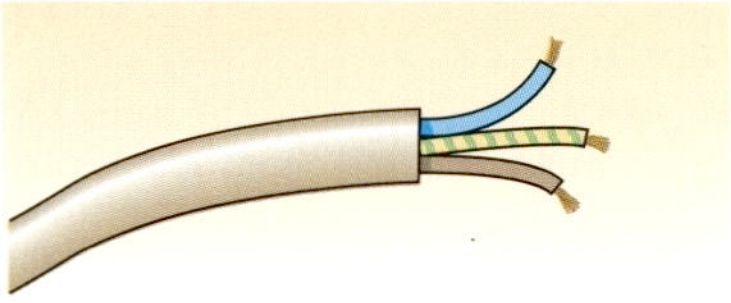

Features of the plug are:

- the case, made of plastic or rubber (good insulators);
- the pins, made of brass (a good conductor);
- a **fuse**;
- an **earth** pin;
- a **cable grip**, to prevent the cable from being pulled out of the plug.

**Did you know?**
A current of 0.1 A can be fatal if it passes through the heart.

The diagram shows how the wires should be connected inside a plug.

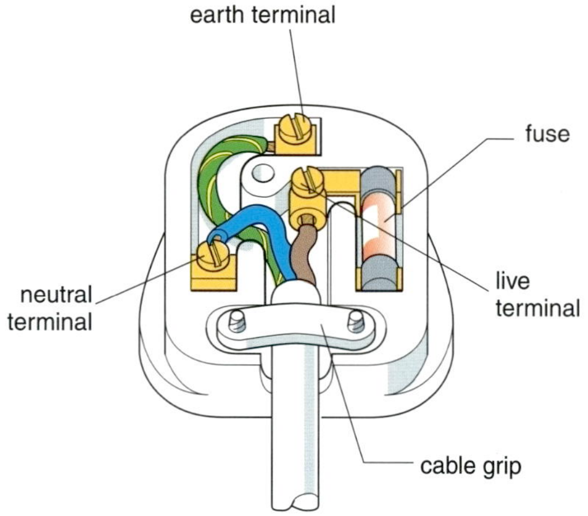


**a** **Which two conducting materials are used in plugs and cables? Which two insulating materials are used?**

**b** **Copy and complete the table to show which wire goes to each terminal.**

| Colour of wire | Name of terminal | Position of terminal |
|---|---|---|
| brown | | right |
| | neutral | left |
| | | top |

## What the wires do

Each of the three wires in a plug has an important, but different, job to do.

- The **live** wire. Energy from the mains electricity supply flows into a house and into an appliance along the live wire. All fuses and switches are connected in series with the live wire.
- The **neutral** wire is needed to make a complete circuit. Just as a torch lamp needs to be connected to both the positive and negative terminals of the battery, a mains lamp needs to be connected to both the live and neutral connections of the mains supply.
- The earth wire is for safety. All appliances with exposed metal parts need an earth wire so that the user cannot get a shock.

**c Which wire to an electrical heater normally has no current in it?**

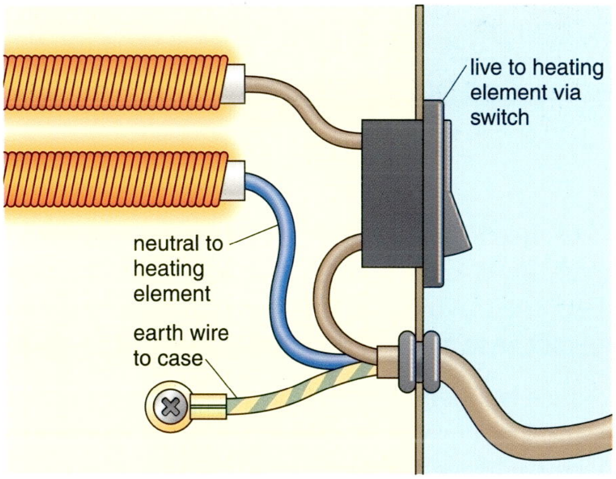

This is how an electrical heater is connected to the mains supply.

A plastic-cased drill, television or hairdryer does not need an earth wire. This is because these appliances are **double insulated**. There are two layers of insulation between the live wire and the casing – the insulation around the wire and the case itself. Even if the live wire becomes loose and touches the inside of the plastic casing, the user does not get a shock because the plastic does not conduct electricity.

All double-insulated appliances carry this symbol.

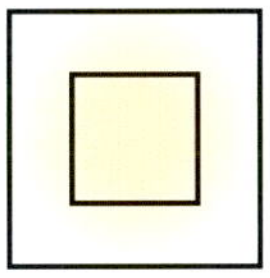

This drill is double insulated. It does not need an earth wire.

## Questions

**1** Which wire includes the fuse?

**2** The drawings show plugs which have been incorrectly wired, or which are dangerous in some other way. Match each drawing to the correct description in the list.

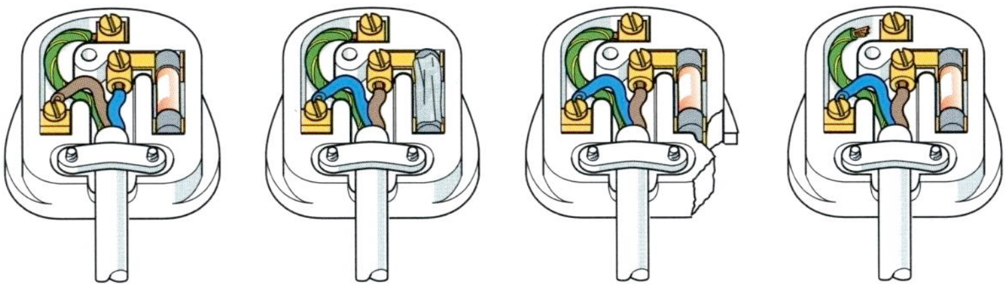

A B C D

**a** Earth wire not connected.

**b** Live and neutral wires swapped over.

**c** Case broken to expose live terminal.

**d** Broken fuse replaced with aluminium foil.

### Key ideas

- *The three wires that connect an appliance to the mains supply are the live, earth and neutral.*
- *A double-insulated appliance does not need an earth wire.*

# 5:6 More safety features

The fuse is needed in a plug to make sure that the current passing in an appliance and its connecting wires does not become too big. Too much current can damage an appliance, or even cause a fire in the appliance or in its connecting wires.

A fuse is labelled with its current rating and contains a thin wire. The current in an appliance also passes in the wire. If the current exceeds the rating of the fuse, the wire overheats and eventually melts. This breaks the circuit.

The current rating of a fuse should be a little more than the current when the appliance is operating normally.

This electric fan caught fire as a result of too much current passing in it.

**a** **Sometimes people hide mains connecting wires under a carpet or rug. Why is this dangerous if a fault causes the wires to overheat?**

**b** **Choose fuses from those in the picture which would be suitable for:**

**a lamp, normal operating current 0.4 A;**

**a heater, normal operating current 8.7 A.**

These fuses are rated at 30 A, 13 A and 3 A.

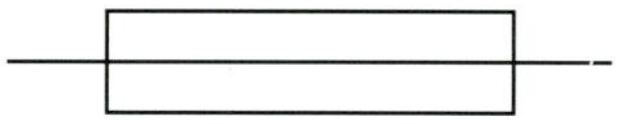

The circuit symbol for a fuse.

## Earthing

Appliances such as washing machines and heaters often have metal cases. These can be hazardous if the live wire makes contact with the case. The case should be connected to the **earth wire**.

If the case now becomes live:

- there is a low resistance current path to earth;
- a large current passes from live to earth;
- this melts the fuse and breaks the circuit.

**No earth connection:**
Touch the live case and you will get a serious shock – possibly fatal. A current of just 0.1 A is enough to kill you.

**Metal case earthed:**
A large current passes from live to earth. The fuse wire melts and breaks the circuit.

**c Explain why a blown fuse should never be replaced with a piece of copper wire.**

## Alternating and direct current

Cells and batteries supply **direct current (d.c.)**. A direct current passes from positive to negative and does not change direction. Mains electricity is **alternating current (a.c.)**. Its direction changes a hundred times each second.

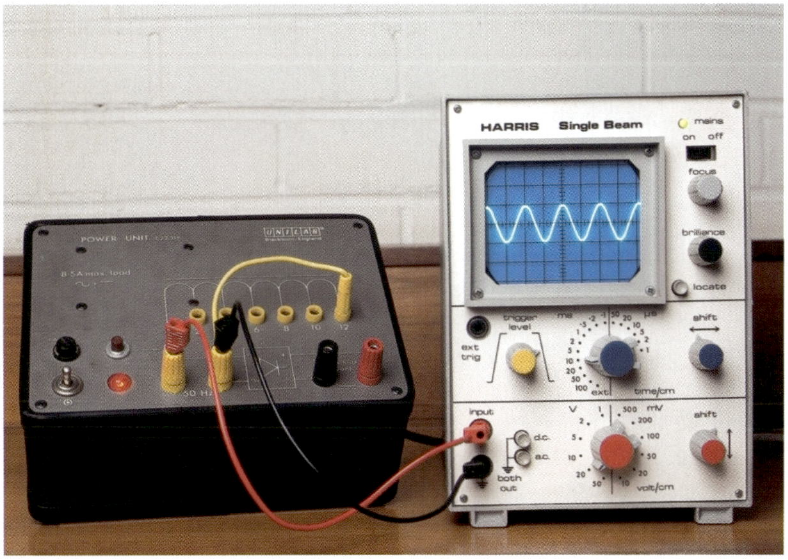

The oscilloscope trace shows that an alternating current changes direction.

**Did you know?**

A cycle dynamo produces an alternating voltage. The faster the dynamo spins, the greater the voltage and the more times it changes direction each second.

An oscilloscope can show the difference between an alternating voltage and a direct voltage:

- If the trace stays above or below the central horizontal line the voltage is not changing direction – it is direct.
- If the trace crosses the central horizontal line the voltage is changing direction – it is alternating.

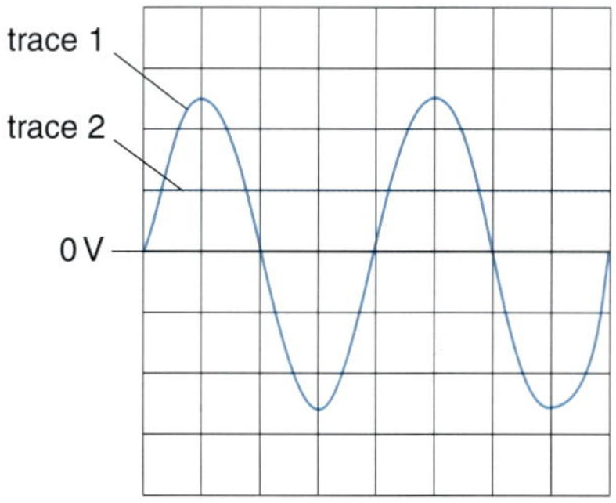


**d Which of the traces shows a direct voltage and which shows an alternating voltage?**

## Questions

**1** Which **two** of the following act together to protect users of an appliance from an electric shock?

earth wire    fuse    live wire    neutral wire

**2** How many wires are used to connect a double-insulated appliance to the mains electricity supply?

**3** What type of electric current is used in a car to start the engine? Explain how you can tell.

## Key ideas

- *Fuses protect against fire hazards.*
- *The fuse in a plug, together with the earth wire, protects against electrocution.*
- *Current from the mains supply is alternating; that from a battery is direct.*

# 5:7 Paying for electricity

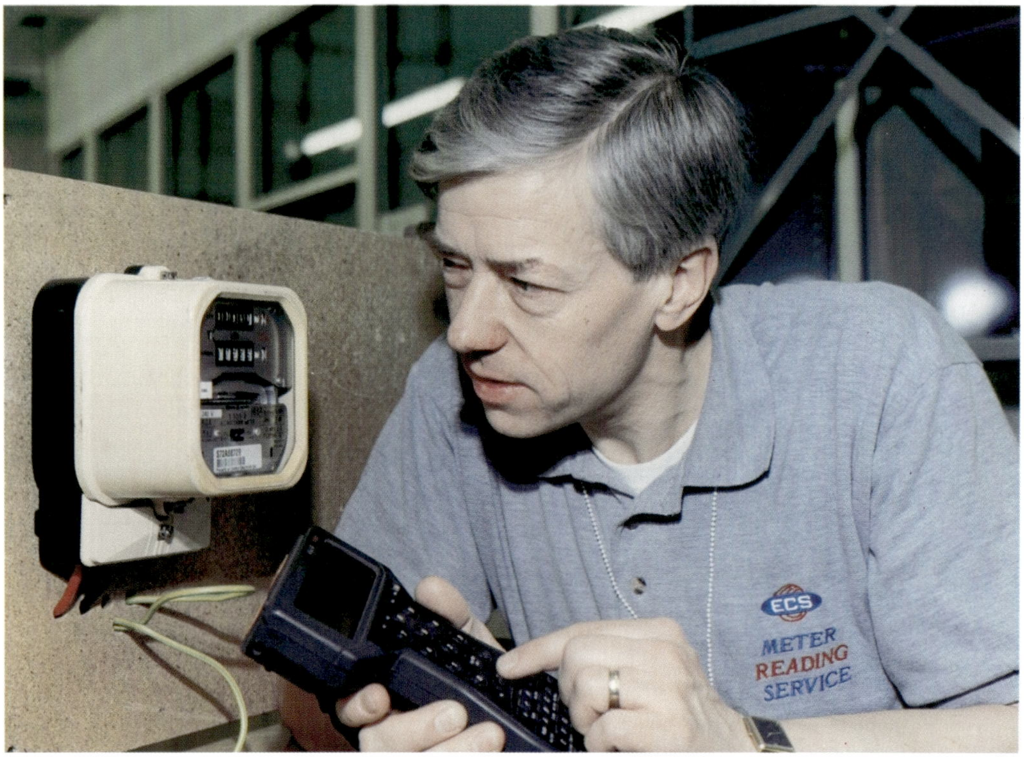


Householders have to pay for the energy they obtain from the electricity supply. The electricity meter measures the amount of energy that flows into the house along the live wire.

The meter reader records the reading on the meter. By comparing this with the previous reading, it is easy to calculate the energy flow between the readings.

## Appliance power

All mains electrical appliances are labelled with their **power** input. A heater could be labelled 1 kW or 2 kW. An electric cooker might be labelled 7 kW. The letters 'kW' stand for '**kilowatt**'. 1 kW is equal to 1000 W.

The power tells you how quickly the appliance transfers energy from the electricity supply.

A 1 kW heater transfers 1000 J each second.

A 2 kW heater transfers 2000 J each second.

**a How much energy does a 7 kW cooker transfer each second?**

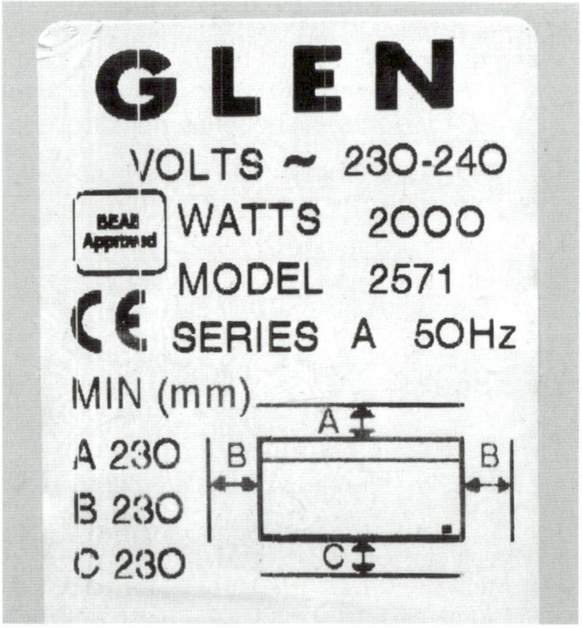


This label is from a 2 kW heater.

## Kilowatt-hours

The joule is a very small unit of energy. If the electricity supply companies measured the amount of energy transferred to a house in joules, they would be dealing with very large numbers. Instead, they use the **kilowatt-hour (kW h)**.

One kilowatt-hour of energy is the energy input to a 1 kW appliance in 1 hour. The actual amount of energy input to an appliance is calculated using the formula:

Energy transferred in kW h = power in kW × time in h

A household electricity bill refers to the 'number of units used'. The **unit** is 1 kW h of energy.

**b How many kilowatt-hours of energy are supplied to a 7 kW cooker transfer in 2 h? How many units is this?**

## Calculating the energy supplied

The reading on an electricity meter is normally taken every three months. The energy supplied is worked out by taking the previous reading away from the new reading.

Present reading = 30 459 kW h

Previous reading = 30 329 kW h

Energy supplied = 130 kW h

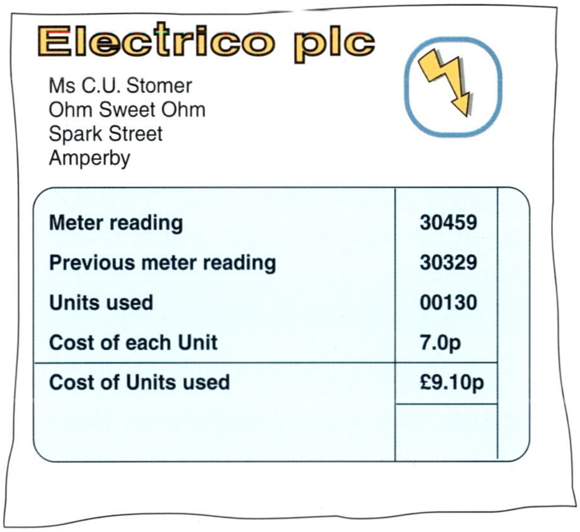

**Electrico plc**

Ms C.U. Stomer
Ohm Sweet Ohm
Spark Street
Amperby

| | |
|---|---|
| Meter reading | 30459 |
| Previous meter reading | 30329 |
| Units used | 00130 |
| Cost of each Unit | 7.0p |
| Cost of Units used | £9.10p |

## Calculating the cost

The electricity bill shows the cost of each unit. This is usually about 7p. It then shows a calculation of the cost of this energy.

Cost of 130 units at 7p each = 130 × 7p = 910p = £9.10.

Here is the formula for calculating the cost of energy from the electricity supply:

Total cost = number of units × cost per unit

To work out the cost of using a single appliance, this formula is written as:

Cost = power in kW × time in h × cost of 1 kW h

**c What is the cost of using a 2.5 kW heater for 5 hours if each kW h of energy costs 7p?**

### Did you know?

Some electricity meters can be read remotely. They send their readings down the cables to the computer of the electricity supply company.

## Questions

**1** A floodlight is labelled 2.5 kW. How much energy does it transfer in one hour?

**2 a** A 3 kW electric kettle is used for 4 hours in a week. How much energy does it transfer in this time?

**b** If each kW h costs 7p, how much does this cost?

**3** The pictures show the readings on the electricity meter in a holiday cottage, recorded at the beginning and end of a week's holiday.

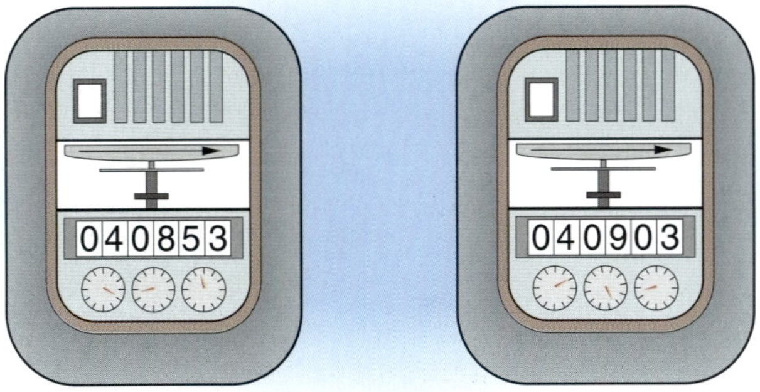


beginning    end

**a** What was the reading at the end of the week?

**b** How much energy passed through the meter during the week?

**c** If each unit costs 7p, what was the total cost of the energy?

### Key ideas

- *1 kW is a rate of energy transfer of 1000 J/s.*
- *A 1 kW appliance transfers 1 kW h of energy in 1 hour.*
- *The cost of using an appliance is calculated as: power in kW × time in h × cost of 1 kW h.*

# 5:8 Electrical heating

We use electricity at home to produce heat, light and movement. It is a very useful source of energy because:

- it is simple to connect to;
- it can be used anywhere in the house;
- it does not leave any waste in our homes;
- it is easily turned on and off.

Whenever a current passes in a wire, there is some transfer of energy from the electricity supply to heat in the wire. How much energy is transferred each second depends on the size of the current and the resistance of the wire.

Mains connecting wires are made out of copper. Thick wires are used so that their resistance is as low as possible. The wire in a filament lamp needs to get hot, so it is made out of a long piece of very thin tungsten wire.

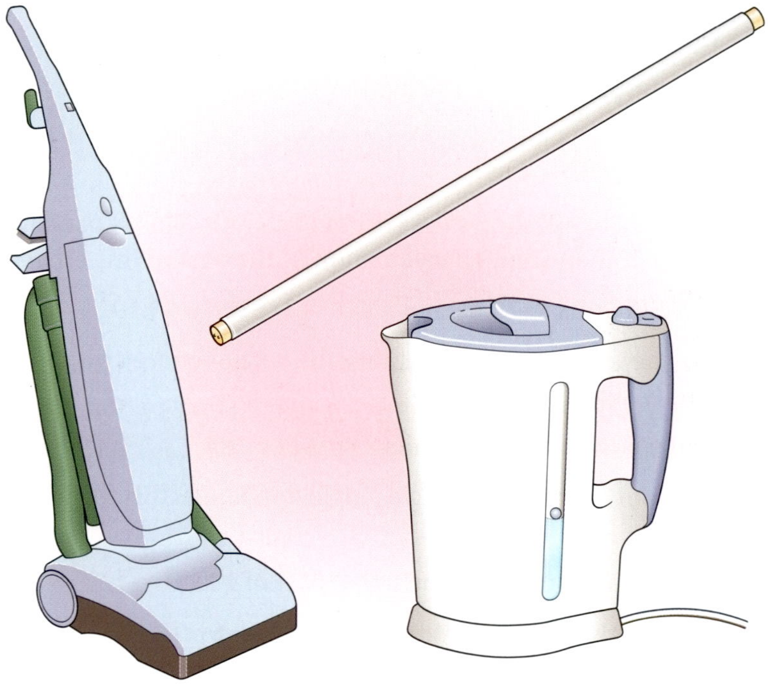

**These appliances use electricity to produce heat, light and movement.**

## Space heating

One of the simplest ways of heating a room is to use an electric bar heater. This gives off energy mainly as **electromagnetic radiation**. Some of this **radiant energy** is light, but most of it is infrared radiation. **Infrared radiation** is similar to light, but it has a longer wavelength and is not detected by our eyes. Like light, it heats any surface that absorbs it.

> **Did you know?**
> A filament lamp operates with the filament at a temperature of about 2000 °C. Tungsten is used for the filament because it has a very high melting point, 3700 °C.

In an electric bar heater:

- Each bar has a power of around 1 kW.
- The heating element operates at a temperature of about 700 °C.
- The element is encased in an insulating material, such as quartz, which can withstand the high temperature and also allow the radiant energy to pass through it.
- A reflector behind the element reflects the radiant energy forwards.

Radiant heaters like the bar heater are very good at heating solid objects, but poor at heating the air in a room. This is because the electromagnetic radiation travels easily through the air with very little of it being absorbed.

**a Why is it an advantage that a radiant heater gives out light as well as infrared radiation?**

## Heating water

Water is a fluid, so it can flow. Energy can move around within it by **convection currents**. When part of the water is heated, it rises above the denser, colder water that surrounds it. The diagram shows the convection current in a kettle that is heating some water.

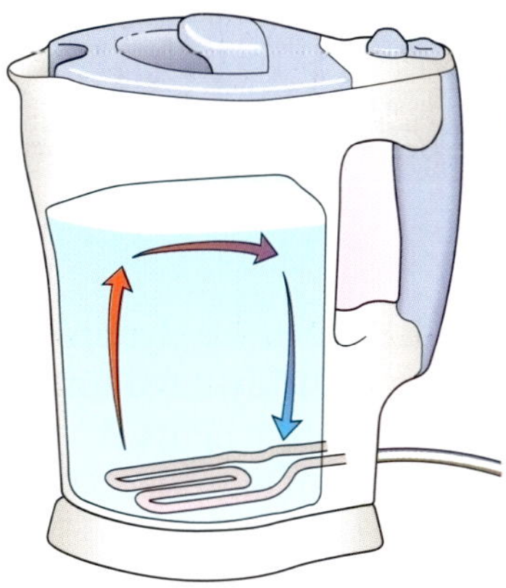

**The heated water rises and is replaced by cooler water.**

The heating element in a kettle:

- has a power of around 2.5 kW;
- is always placed at the bottom of the kettle;
- is in a metal case that transfers energy to the water by **conduction**.

**Immersion heaters** work in a similar way to a kettle. When it is to be used just occasionally, the heating element is fitted at the top of the hot water tank. This has the advantage that it heats only the water at the top of the tank, so it heats up quickly.

A household that relies on immersion heating for its hot water supply normally uses two heaters – one at the top of the tank and one at the bottom.

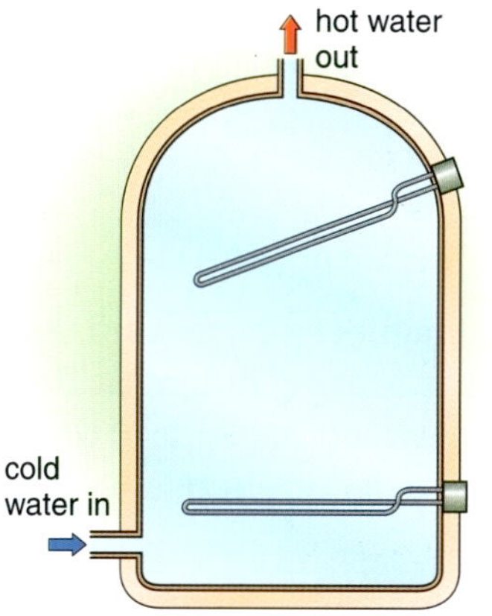


**This tank uses two immersion heaters.**

**b Which heater only heats the water at the top of the tank?**

**c Which heater heats all the water in the tank?**

## Smoothing things out

The heating element in an iron is inside the metal plate. This element:

- has a power of around 1500 W;
- transfers energy to the clothes by conduction through the metal plate;
- has a **thermostat** so that it maintains a steady, but variable, temperature;
- is also used in some irons to produce jets of steam.

**d Why is at an advantage to be able to vary the temperature of the iron?**

## Questions

**1** Copy out the sentence, using words from the list to fill in the blank spaces.

**gamma infrared light radiant**

An electric bar heater emits energy mainly in the form of __________ and __________ radiation.

**2** Why should the wires that connect a bar heater to the mains supply have less resistance than the heating element?

**3** Immersion heaters are fitted with thermostats set to a temperature of 50 °C.

Explain why it would be dangerous for the water in a hot water tank to be allowed to boil.

### Key ideas

- *A bar heater emits radiant energy.*
- *Kettles and immersion heaters rely on convection currents to heat all the water.*
- *Energy is transferred through the metal plate of an iron by conduction.*

# Generating electricity

If you have a bicycle, it may have a **dynamo** fitted. As you travel along, the dynamo turns and generates electricity for the lights. A dynamo saves you the cost of batteries, and it never runs out.

**a** What is the disadvantage of a dynamo:

**i** when you are travelling slowly

**ii** when you stop at traffic lights?

As the wheel spins around, it spins the magnet.

## A moving magnet

Inside a dynamo, a magnet spins around. As it spins past the coil, the moving **magnetic field** produces a voltage between the ends of the coil. This voltage makes a current pass in the lights.

Generating electricity like this is called **electromagnetic induction**.

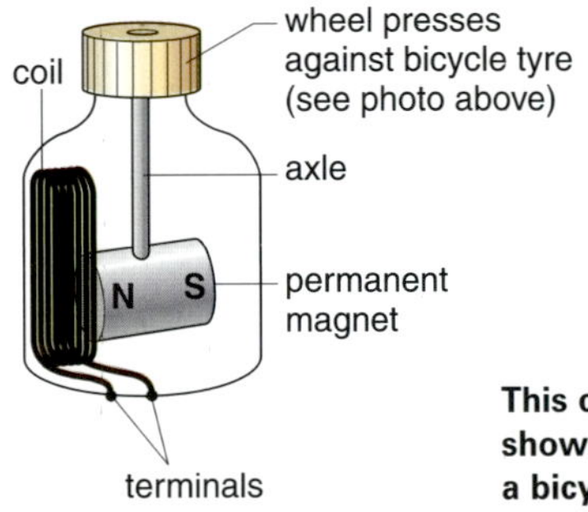


This cut-out view shows the inside of a bicycle dynamo.

## Inside a power station

The **generators** in a power station are large-scale versions of a bicycle dynamo. There are some important differences:

- The magnets that spin are **electromagnets**. These have their own small generators, which you can see to the right of each of the main generators in the photograph.
- The electricity is generated at a very high voltage, about 25 000 V, and a very high current, about 20 000 A.
- The electricity is generated in thick copper conductors that surround the electromagnet. These conductors are cooled by water flowing on the inside and by hydrogen gas flowing over the outside.

Each of these large red generators in a power station produces enough electricity to supply a small town.

## Investigating electricity generation

Electricity is generated in a conductor when the magnetic field passing through it *changes*. The diagram shows how you can investigate this, using a magnet and a coil of wire connected to a sensitive ammeter.

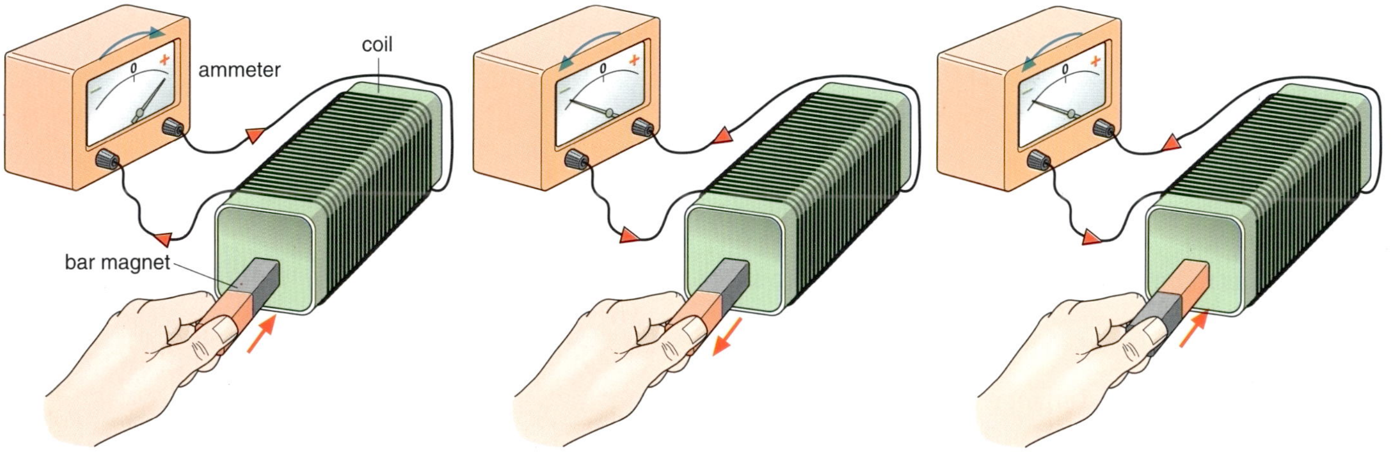


These experiments show that:

- The direction of the current depends on which pole points towards the coil and which way the magnet is moved.
- The faster the movement, the bigger the current.

**b State two ways of changing the direction of the current in the coil.**

**Did you know?**
The first power station was opened at Godalming in Surrey in 1881. It was powered by a water wheel.

## A bigger voltage

Experiments like these show that there are four ways of *increasing* the voltage caused by a magnet spinning inside a coil:

- spin the magnet faster;
- use a stronger magnetic field;
- use a coil with more turns of wire;
- use a coil with a greater area.

**c List four ways of *decreasing* the voltage caused by a magnet spinning inside a coil.**

## Questions

**1** Copy and complete the following sentences, using words from the list to fill the gaps.

**current voltage magnet coil circuit**

In a bicycle dynamo, a _________ spins around inside a _________.

A _________ is produced between the ends of the coil. This causes a _________ to pass in the lights.

**2** State three differences between a power station generator and a bicycle dynamo.

## Key ideas

- *Electricity is generated on a small scale by a magnet spinning near or inside a coil of wire.*
- *In a power station electricity is generated by electromagnets spinning inside thick copper windings.*

# 5:10 Electricity nationwide

Giant **pylons** hold up the **power lines** that carry electricity around the country. The electricity is at a very high voltage, sometimes 400 000 V. This is very dangerous, so the pylons keep the wires safely above our heads.

These power lines carry electricity from power stations to where it is needed. They make up the National Grid.

## Changing the voltage

Electricity has to be transmitted at a high voltage to keep the current as low as possible. The greater the current, the more energy is wasted in heating up the wires.

**Transformers** are used to change the voltage of the electricity supply. There are transformers in sub-stations near your house and school. Signs warn of the danger of the high-voltage electricity.

**a** **Where have you seen electricity sub-stations like this? What warning signs did they have?**

This electricity sub-station has transformers that reduce the voltage to the level needed by the consumers. People have been killed when they have entered sub-stations like this.

## Around the country

Electricity is generated in power stations. Then it must be distributed around the country. **Step-up transformers** are used to increase the voltage at the power station. Then **step-down transformers** reduce it again before the electricity goes to places such as houses and factories.

**b** **Look at the diagram that shows part of the National Grid. What is the voltage of the electricity generated in the power station?**

At what voltage is it transmitted along the power lines?

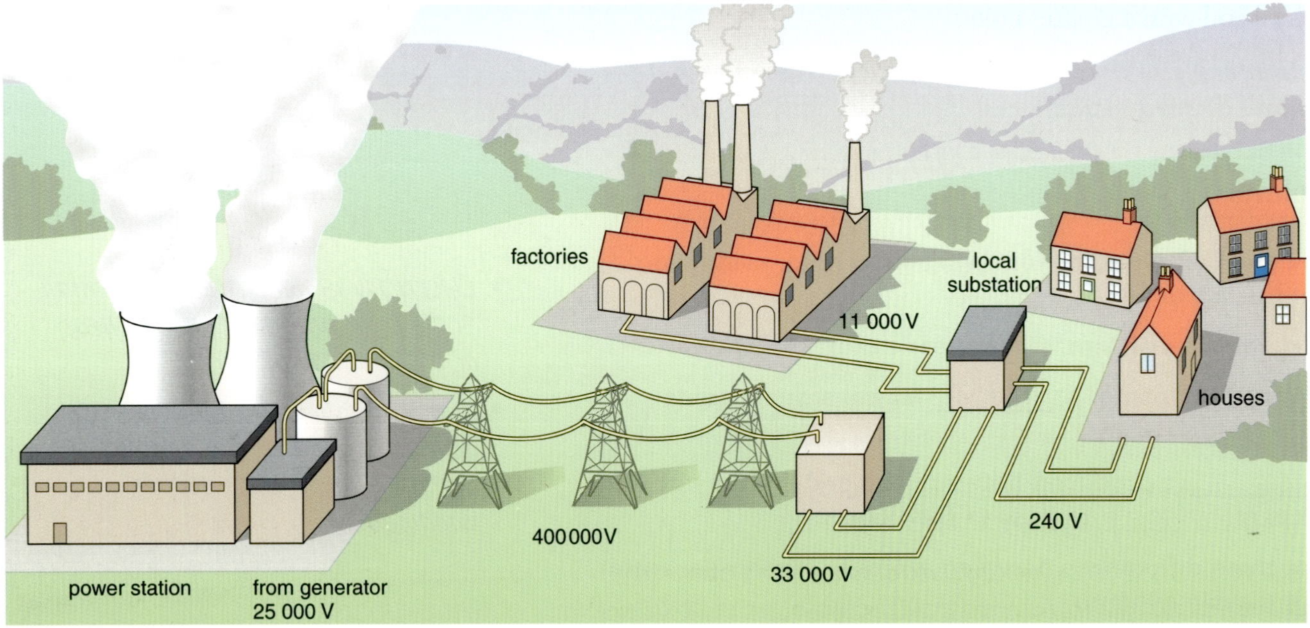

## Overhead or underground?

Overhead cables, supported by pylons, are used to transmit electricity across the country. Overhead cables are ugly and more prone to damage during severe weather than underground cables are, but they do have several advantages.

- The air gap between the cables and the ground acts as a good electrical insulator.
- The air surrounding the cables acts as a coolant, removing excess heat by convection currents.
- They are cheap to install and running costs are low.
- Faults are easily spotted, using a helicopter.

The cables used to take electricity from substations to houses are buried underground. These do not carry very high voltages and currents, unlike those that make up the National Grid, so they can be insulated and buried safely.

## Going under

In towns and cities and near buildings, overhead cables on pylons would be dangerous, so even those carrying very high voltages and currents need to be buried.

These cables are expensive to install and maintain because:

- high voltages need very good electrical insulation;
- good electrical insulation is also good heat insulation;
- the cables have to be cooled by pumping oil through them.

**c State two advantages of using underground cables rather than overhead ones.**

**Did you know?**
In France, some power lines carry electricity at over 1 million volts.

## Questions

1 Why are hazard-warning signs displayed on electricity sub-stations?

2 Copy the boxes below. Choose words from the list to show how electricity flows from the power station to the user. You will have to use one of the words twice.

**transformer generator consumer power lines**

3 Copy the sentences below, choosing the correct word from each pair.

At a power station, transformers increase/decrease the voltage of the supply before it is transmitted along the power lines.

At a sub-station, transformers increase/decrease the voltage of the supply before it reaches the consumer.

4 Write a letter to a local newspaper complaining about the proposed building of a new power line across the countryside.

## Key ideas

- *Transformers are used to make the voltage of the electricity generated at the power station bigger.*
- *Transformers in sub-stations make the voltage smaller.*
- *Overhead cables are cheaper to install than underground ones, and they have lower running costs.*

# 5:11 Energy from fossil fuels

Every time you use an electrical device at home, energy flows from a power station to your home.

This giant power station produces over a million kilowatt-hours of electricity every hour. That's enough electricity for quite a big city. In the picture you can see:

- the store of coal, waiting to be burned;
- the chimney, which releases waste gases from the burning coal;
- the cooling towers which pass waste heat into the atmosphere.

## Inside the power station

Burning coal is used to heat water. The steam produced by the **boiler** is very hot – over 700 °C. It is also at a very high pressure. It rushes along pipes to the turbines.

The moving steam supplies energy to the turbines to keep them turning.

The turbines turn the electromagnets in the generators. Thick cables carry the electricity away from the generators.

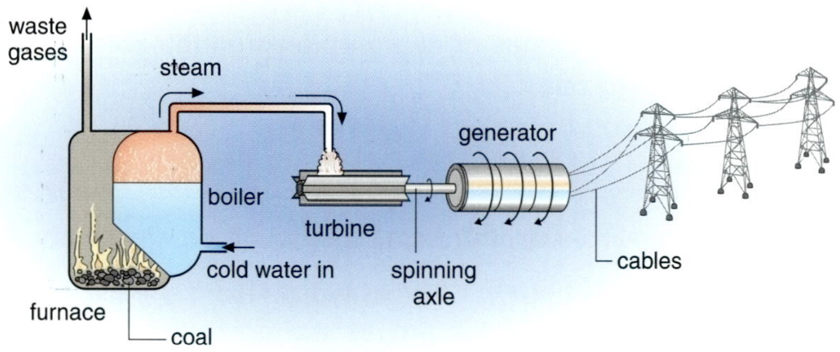

How electricity is generated at a coal-fired power station.

**a Which machine produces the electricity in a power station?**

## Fossil fuels

Electricity has helped to make our lives easier, but it's not all good news.

Most power stations in the UK burn coal or gas. These are **fossil fuels**.

Fossil fuels were formed from the remains of living plants and animals that died millions of years ago. They are stores of energy, trapped from the Sun when they were alive. Once they have been burned, we cannot replace them. They are **non-renewable** energy stores, and one day we may run out of them.

**Did you know?**
France has no fossil fuels. It relies on nuclear power to generate electricity.

## The greenhouse effect

When fossil fuels are burned, carbon dioxide is released into the atmosphere. Coal produces the most carbon dioxide.

Extra carbon dioxide in the atmosphere is a problem. This increases the **greenhouse effect**, making the Earth hotter.

There is evidence that the Earth's average temperature is rising. It may be affecting our climate, making it wetter. It may also make sea levels rise, swamping some low-lying parts of the world.

**b Which produces the least carbon dioxide when it burns: coal, oil or gas?**

## Acid rain

Coal and oil often contain sulphur. When this burns, it becomes sulphur dioxide. This is released into the atmosphere from the power station chimney.

Sulphur dioxide is acidic. When it dissolves in rain, the result is **acid rain**, which pollutes the environment. Plants die; fish and other water creatures are poisoned.

There are two ways to avoid this:

- remove the sulphur before burning the fuel;
- remove the sulphur dioxide from the power station's waste gases.

Both of these are expensive, and add to the cost of producing electricity by burning coal.

**c What substance is produced when sulphur burns? Is it acidic or alkaline?**

These trees have died because acid rain has poisoned the ground in which they are growing.

## Questions

**1** Name three fossil fuels.

**2** Copy and complete the diagram, which shows how a fossil fuel power station works.

**3** When fossil fuels are burned, polluting gases are produced. Which gas:

**a** increases the greenhouse effect

**b** contributes to acid rain?

**4** Put these fuels in order, starting with the one which produces most carbon dioxide:

coal gas oil

### Key ideas

- *Coal, oil and gas are fossil fuels.*
- *When fossil fuels are burned, gases are released that can pollute the environment.*
- *Carbon dioxide increases the greenhouse effect and adds to global warming.*
- *Sulphur dioxide causes acid rain.*
- *Coal produces the most carbon dioxide; natural gas produces the least.*

# 5:12 Energy from wind and water

Fossil fuels and nuclear fuels are non-renewable. We must look for other, **renewable energy** sources to supply our demand.

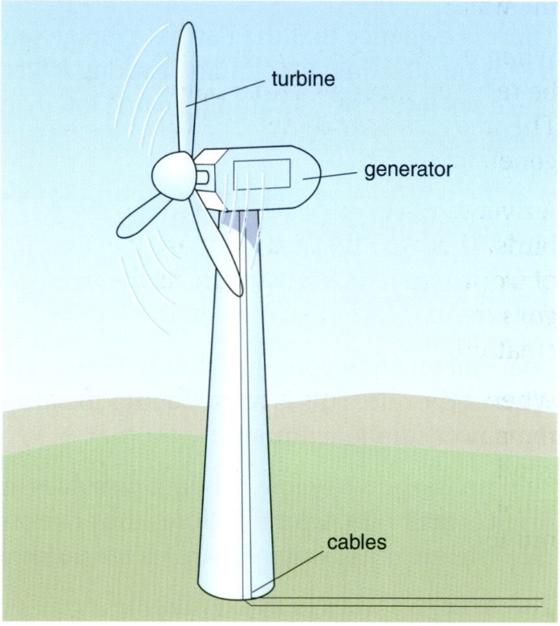


## Moving air

The wind is moving air. It has **kinetic energy**. This energy can be used to produce electricity by a wind generator. The wind will always blow. It will never run out so it is a renewable energy source.

The large blades of a **wind turbine** are driven round by the wind. The turbine drives a generator, producing electricity.

**a What energy transfer takes place at a wind generator?**

## Wind farms

Wind generators are built in windy places:

- on hillsides;
- on the coast.

Several generators together make up a **wind farm**. If you are close to a generator, you may hear its blades whizzing round. If you live nearby, you may consider this a form of noise pollution. A wind farm may spoil the look of the landscape – a form of visual pollution.

**b Which part of a wind generator is the turbine?**

**Did you know?**
In the future wind farms may be built offshore to provide some of our electricity.

## Hydroelectricity

A tiny amount of the electricity generated in the UK comes from moving water. This is **hydroelectric power**. Some streams and rivers in Scotland and Wales flow fast enough to drive water turbines linked to generators. In some countries large dams are used.

The giant dam in the photograph is storing water for a hydroelectric scheme.

- The water behind the dam stores a lot of gravitational potential energy.
- As the water falls down through pipes, this energy is transferred to kinetic energy.
- The kinetic energy of the water is used to keep the turbine blades going round, generating electricity.

**c Explain why it is not possible to remove all the kinetic energy from the moving water.**

The Hoover Dam on the Colorado River, USA generates more than I million kW of electricity.

## Energy from the tides

A river estuary is a good place to build a **tidal power** scheme. When the tide comes in, the estuary fills with water. A barrage (a dam) can be built across the estuary to trap the water.

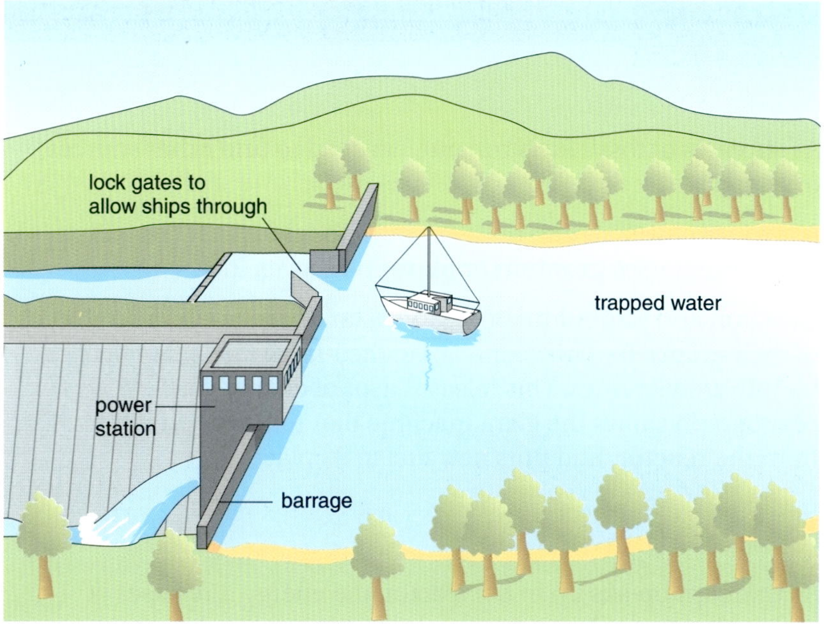


When the tide goes out again, the water can be released through a tidal power station. The water turns turbines that turn generators.

A river estuary is a good place for wading birds. There are mudflats that are the habitat of worms and other creatures. As the tide goes out, the birds can feed on these creatures.

When a tidal barrage is built, the mudflats remain underwater. They are no longer a good source of food for the birds, which must move away. This means that the number of wading birds declines.

## Energy from waves

Waves are moving water. In a **wave-powered** generator, as the waves go up and down, they make a turbine spin. The turbine drives a generator.

So far, attempts to generate electricity from waves have proved very costly. As the generators float on the sea, they are prone to damage during severe weather conditions.

## Questions

1 Give **two** ways in which a wind farm may damage the environment.

2 Why is the coast a good place to build a wind farm?

3 Choose words from the list to answer the questions which follow.

**reactor generator cables turbine**

a Which part of a wind generator is made to turn round by the wind?

b Which part produces electricity?

4 Here are some people who might be affected when a new hydroelectric scheme is built:

| farmer | angler (fisherman) | forestry worker | windsurfer |
|---|---|---|---|

a Who might be harmed by the new scheme?

b Who might benefit from the new scheme?

### Key ideas

- *Electricity can be generated from the energy in moving air and water.*
- *These are renewable forms of energy.*

# 5:13 Nuclear power stations

As supplies of fossil fuels run out, we need to find other sources of energy.

A nuclear power station does not burn fossil fuels. It contains a **reactor** that uses **uranium** or **plutonium** as its fuel.

Uranium and plutonium both store a lot of energy in the nuclei of their atoms. In a nuclear reactor, their large nuclei are broken up into smaller ones. This releases a lot of energy. The photograph shows the giant machine that removes used fuel from the **reactor**, and puts new fuel in its place.

## Inside the power station

As uranium breaks up in the reactor, the energy it releases is used to heat a coolant such as water. This is used to generate high-pressure steam, just as in a coal-fired power station.

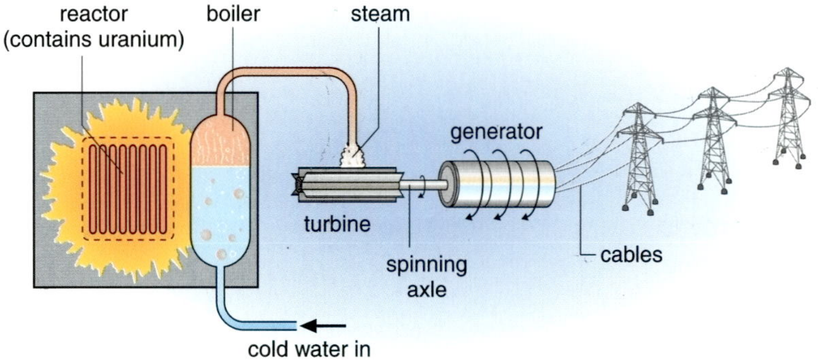


**a Which part of the nuclear power station contains the uranium fuel?**

## Radioactive decay

In a nuclear power station, the uranium fuel does not burn. No waste gases are produced.

As the uranium breaks up into smaller nuclei, it releases energy as heat. It also produces **radioactive** waste products, which are very hazardous.

In normal operation, very little **radiation** or radioactive material escapes from the power station. However, if there is a serious accident, large amounts of radioactive material may escape. It can be spread over large areas.

This nuclear power station at Chernobyl in the Ukraine, exploded in 1987. Radioactive material spread over several villages and large areas of farmland. Many deaths resulted.

**b Which of the following are produced when a nuclear fuel such as uranium decays: radiation, carbon dioxide, radioactive substances, sulphur dioxide?**

## Radioactive waste

The waste products are hazardous because of the radiation they produce. They keep on emitting radiation for thousands of years.

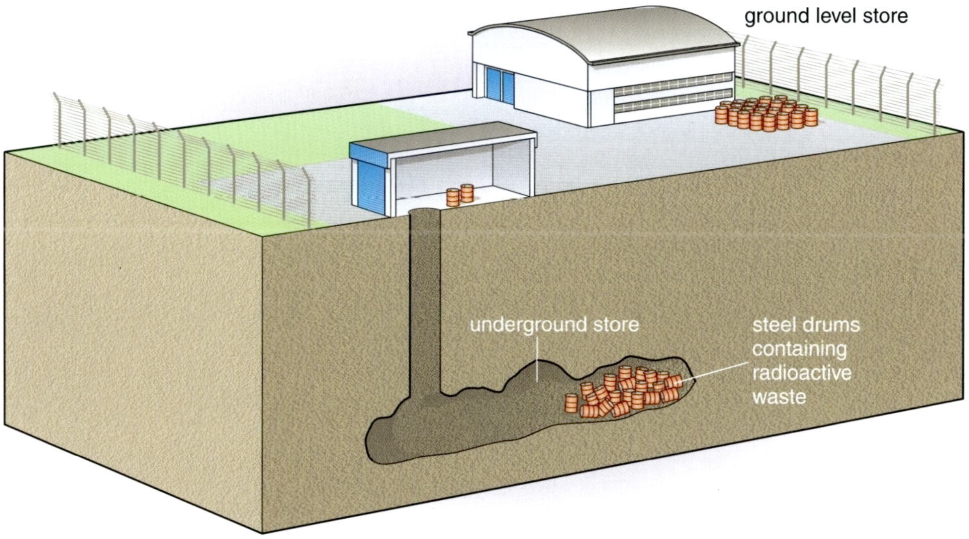


There are two ways of storing **radioactive waste**:

- bury the waste deep underground;
- store it in steel containers at ground level, where it can be carefully monitored.

The radiation from radioactive substances can cause cancer. The more of this radiation someone is exposed to, the greater their chance of contracting cancer.

Uranium comes from mines. Eventually, we may use up all the available stores of uranium. It is a non-renewable fuel.

**c Why must waste from a nuclear power station be stored carefully for a very long time?**

**Did you know?**
Nuclear power adds only 0.1% to the background radiation – far less than that due to the use of radioactive materials in medicine.

## Questions

1 Which part of the uranium atom releases energy when it is broken up?

2 The following sentences explain how electricity is generated in a nuclear power station. Copy them out, choosing the correct word from the list to fill each gap.

**generator boiler reactor turbine**

Uranium fuel in the _________ releases a lot of energy.

This energy heats water to produce steam in the _________.

The steam makes the _________ spin around.

This turns the _________ which produces electricity.

3 a In what way is radioactive waste a health hazard?

b How can this hazard be reduced?

## Key ideas

- *Nuclear power stations use uranium or plutonium as their fuel.*
- *Nuclear fuels do not produce the same polluting gases as fossil fuels.*
- *In an accident, large amounts of radioactive material may be spread over a wide area.*
- *The waste produced by a nuclear power station can stay dangerously radioactive for thousands of years, and must be stored safely.*

# 5:14 Using energy from the Sun

The satellite in the photograph uses **solar cells** to charge its batteries. It can turn the large panels of cells so that they are always facing the Sun.

A solar cell produces electricity directly from sunlight. Most of the energy of sunlight ends up as heat, as it warms the solar cells, but a small amount is transferred to electricity.

**a Why is it an advantage to be able to turn the panels of cells?**

## On a sunny day

We can also use solar cells on Earth. They work best in bright sunlight. At night they produce no electricity. The brighter the light falling on them, the more electricity they produce.

Some modern buildings have solar cells on the outside walls and roof. Large batteries are needed to store the energy for use at night.

**b At what time of year do solar cells produce the most electricity?**

**At what time of year is the demand for electricity the greatest?**

## Expensive electricity

Sunlight, the energy source for solar cells, is free. However, at present, solar cells are expensive. This means that, over the lifetime of a solar cell, the cost of the electricity it produces is high.

People have to think carefully about where to use solar cells. They are useful in remote places where it is difficult to supply electricity by any other means.

Space is a remote place. That is why many satellites use solar cells. In remote parts of Australia, roadside phones are powered by solar cells. This is much cheaper than laying cables for hundreds of kilometres.

Solar cells are also useful where only a little electricity is needed. That is why some watches and calculators use solar cells.

**c How is energy stored so that the telephone can be used after dark?**

This solar-powered phone box is in the outback in Australia.

## Solar heating

The photograph shows some **solar heating** panels that are used to heat the water for a swimming pool.

Such panels are a common sight on rooftops in southern Europe, where many houses use them for their hot water supply.

- The panels are black so that they absorb as much energy from the Sun as possible.
- This energy is used to heat water that flows through the panels by convection currents.
- The hot water is stored in a tank.

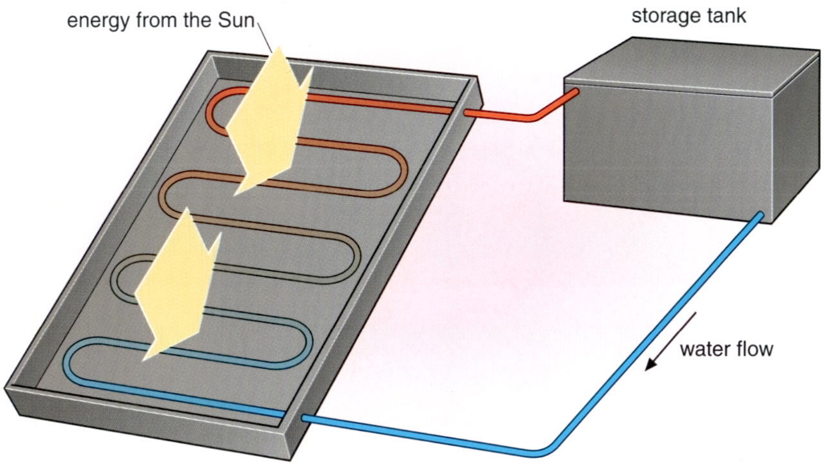


> **Did you know?**
> In the Solar Challenge race, solar-powered cars race over two thousand miles across Australia, from Darwin to Adelaide.

## Questions

1 Copy and label this energy flow diagram which shows how a solar cell transfers energy.

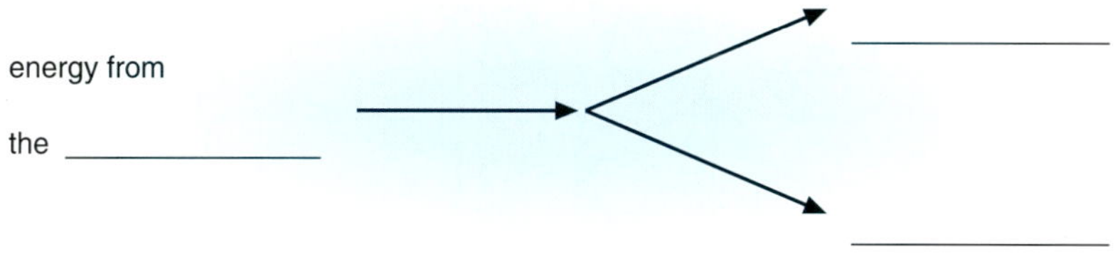


2 An emergency telephone is sited on a high mountain pass, for use by walkers when they are in trouble. Why would solar cells be a good choice to provide the electricity for this phone?

3 a Explain why solar heating panels are usually tilted.

b i Which way does water flow through a solar heating panel?

ii Explain why the water flows in this direction.

## Key ideas

- *Solar cells produce electricity directly from sunlight.*
- *Solar cells are expensive. This means the electricity they produce is expensive.*
- *Solar cells are useful in remote places (e.g. on satellites) and in watches and calculators.*
- *Solar heating panels use energy from the Sun to heat water directly.*

# 5:15 Cutting the cost

The chefs in the photograph are working in a very hot kitchen. A lot of energy is supplied to the cookers and grills. Only a little is used in cooking the food – the rest escapes to the surroundings, making the kitchen a very hot place to work.

A microwave oven makes better use of the energy supplied to it. Most of the energy supplied to it is transferred to heat in the food. Only a little energy escapes – if you put your hand inside the oven after it has been used, you will find that it is not very warm.

This means that a microwave oven wastes less energy than an electric or gas oven.

**a Which needs to be warmed up before being used to cook food: a microwave oven or a gas oven?**

## Efficient appliances

We want the electrical appliances we use to make good use of the energy that is supplied to them. Electricity costs money and generating electricity uses up our fossil fuel reserves, so we don't want to waste it.

A microwave oven transfers a lot of the energy supplied to it to the food, more than a gas or electric oven do. The microwave oven is more **efficient** than the other ovens.

To decide how efficient a device (such as an oven) is, two things need to be considered:

- How much energy is supplied to the device?
- How much of this energy is transferred to where it is wanted?

An efficient device wastes less energy than an **inefficient** one.

**b An electric oven has an input power of 2.8 kW and a microwave oven has an input power of 0.85 kW. Give two reasons why the electric oven needs a higher input power than the microwave oven.**

## Energy-efficient lighting

An ordinary filament lamp is much cheaper to buy than an **energy-efficient** lamp. However, it does not last as long and it wastes a lot more energy. Only about 5% of the energy supplied to a filament lamp ends up as light.

**c What happens to the rest of the energy supplied to the lamp?**

The table compares two different lamps. One is much more efficient than the other.

> **Did you know?**
> Over the lifetime of an energy-efficient lamp the saving in the cost of electricity is £30 – far more than the extra cost of buying it.

| | | Amount of energy supplied each second | Amount of energy transferred as light each second |
|---|---|---|---|
|  | Ordinary filament lamp | 100 J | 6 J |
|  | Low-energy lamp | 20 J | 6 J |

From the last column of the table, you can see that both lamps produce the same amount of light each second.

From the middle column, you can see that the filament lamp must be supplied with 100 J of energy each second. The low energy lamp requires only 20 J each second.

**d The two lamps are of equal brightness. How can you tell this?**

**e Which lamp is more expensive to run? Explain your answer.**

## Questions

1 Which lamp gives off more heat: a filament lamp or a low-energy lamp?

2 A new car is more efficient than the old one it replaces. Does this mean that the new car wastes more energy than the old one, or that it wastes less energy?

3 An electric train wastes 30% of the energy supplied to it. A diesel train wastes 60%. Which is more efficient?

4 What are the benefits of using energy-efficient electrical appliances:

a to a householder

b to the planet?

### Key ideas

- *All devices waste some of the energy supplied to them.*
- *An efficient device wastes a smaller fraction of the energy supplied to it than an inefficient device does.*

# 5:16 Insulating a house

If you could keep all the warmth in your house, you would save a lot of money. When it is warmer inside than outside, energy flows out through the walls, floor, windows and roof. More escapes every time you open a window or a door.

## Saving energy

The picture below shows a house before and after it has been fitted with insulation. The figures tell you how much energy escaped from the house each second on a cold day, before and after the insulation was fitted.

How much energy is saved by insulating the cavity between the inner and outer wall?

Energy lost each second before fitting insulation = 2500 J

Energy lost each second after fitting insulation = 1000 J

So, energy saved each second by fitting insulation = 1500 J

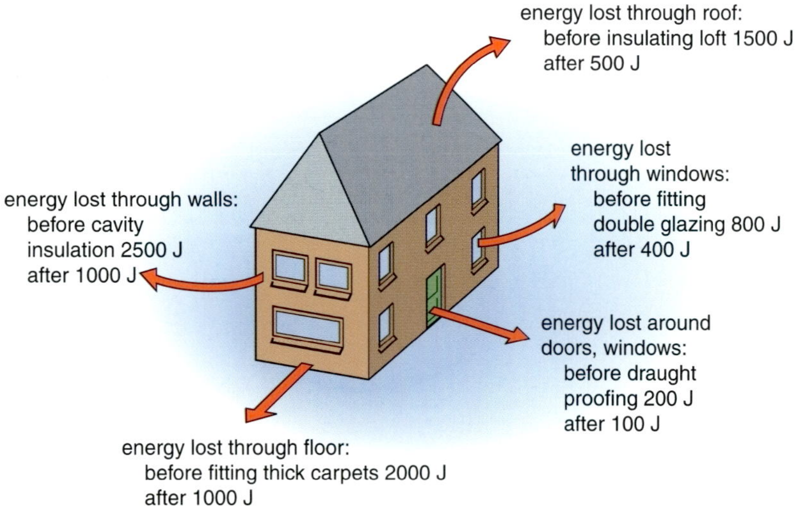


**a** **Work out how much energy is saved by fitting insulation in the loft.**

**b** **Which saves more energy: cavity wall insulation or loft insulation?**

## Saving money

Most new houses are fitted with **double glazing**. This is good because it saves energy, stops draughts and cuts down on noise. However, it is expensive to fit double glazing to an old house, and it doesn't save much money. It is not very **cost effective**.

If you live in an old house, it could be better if you fit cavity wall insulation. Here are some figures to show how to decide whether cavity wall insulation is cost effective.

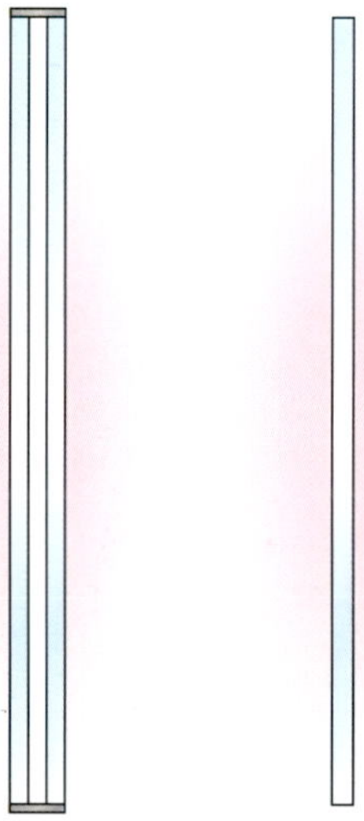

**A double-glazed window and a single-glazed window.**

Cost of cavity wall insulation = £600

Saving on heating bill in one year = £150

So, after 4 years, you will have saved the cost of the insulation. After that, you will save £150 every year. The **pay-back period** is 4 years.

| Method of saving energy | Initial cost | Saving each year | Cost-effectiveness |
|---|---|---|---|
| cavity wall insulation | quite expensive | a lot | good value |
| loft insulation | quite cheap | a lot | very good value |
| thick carpets in all downstairs rooms | expensive | a lot | quite good value |
| draught proof doors | cheap | not much | very good value |
| double glaze windows | expensive | not much | poor value |

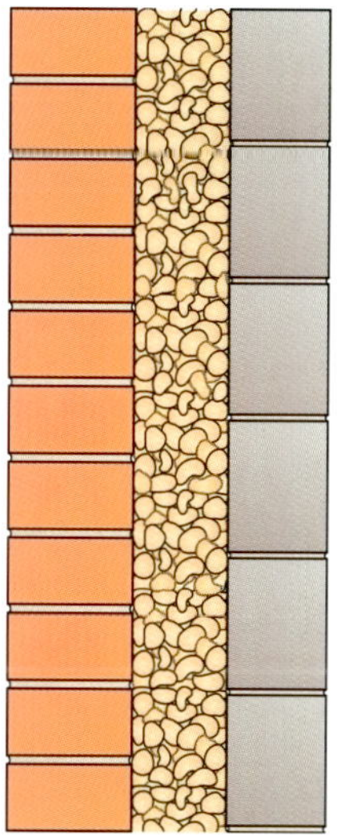

This shows the insulation in the cavity between the inner and outer walls of a house.

**c Which method of saving energy is the least cost-effective? Which are the most?**

The initial cost of saving energy can be high. Eventually, you will save more money than you have spent. If you do not have much money, you may not be able to afford the initial cost. Then your energy bills will remain high.

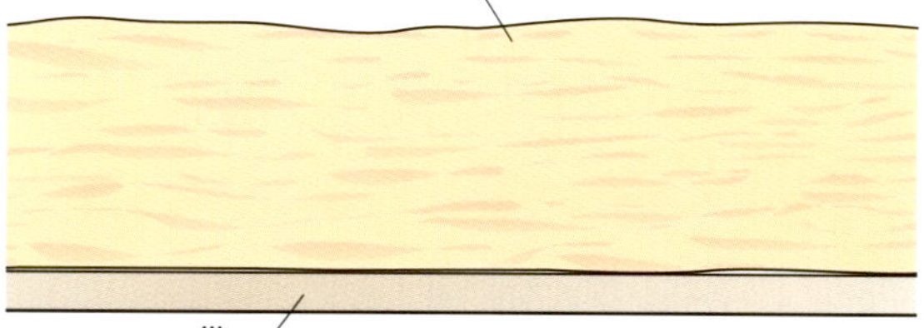


Loft insulation reduces heat loss through the ceiling.

## Thinking of the future

Saving energy saves money. There are other reasons for using energy more efficiently.

- We want to conserve our reserves of fossil fuels so that they last as long as possible.
- Burning fuels causes pollution that damages the environment.

## Questions

**1** Look at the pictures of the house on the opposite page.

**a** Before insulation is fitted, how much energy escapes each second?

**b** How much energy escapes each second after insulation is fitted?

**c** How much energy is saved each second by fitting insulation?

**2** Give three reasons for fitting double glazing. Why are many people reluctant to fit it?

**3** The thermostat controls the temperature of a house which has central heating. You can save more money by turning down the thermostat. Explain why less energy escapes when the thermostat is turned down.

## Key ideas

- *Insulation reduces the flow of energy from a warm object to a cooler one.*
- *It is important to save energy to reduce costs and to help the environment.*
- *Loft insulation, cavity wall insulation and double glazing are methods of insulating a house.*

# End of module questions

## Low demand

**1** A voltmeter measures

**A** current in amps

**B** current in volts

**C** voltage in amps

**D** voltage in volts

**2** Which meter in the circuit is in the correct position to measure the voltage across the lamp?

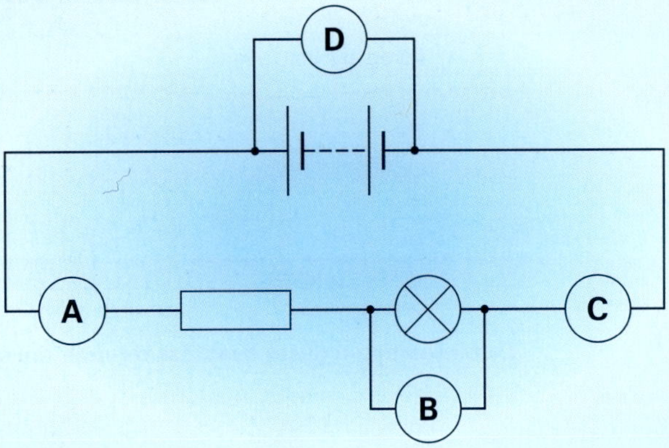


**3** The current in a series circuit can be increased by

**A** adding a resistor in series

**B** increasing the voltage of the supply

**C** decreasing the voltage of the supply

**D** adding a voltmeter in series

**4** Which graph shows how the resistance of a light-dependent resistor (LDR) depends on the intensity of the light shining on it?

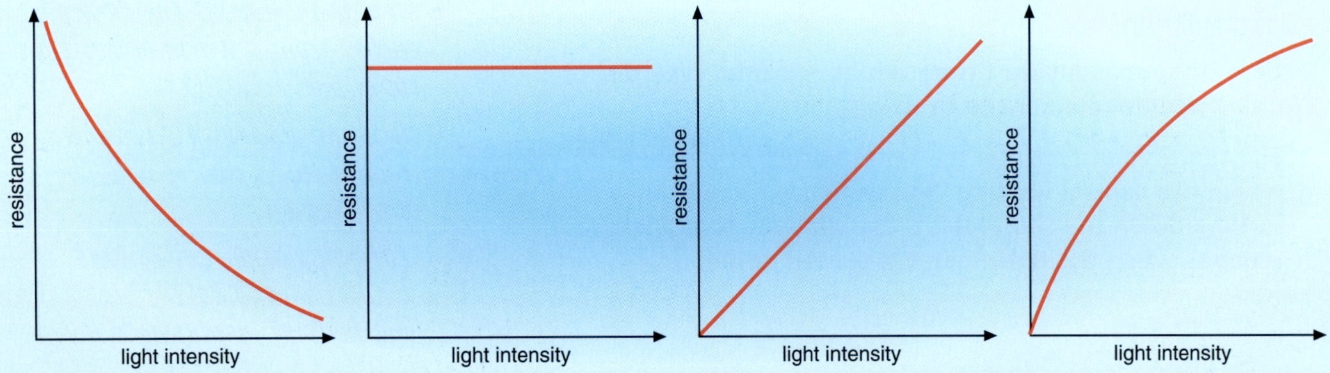


**A** **B** **C** **D**

**5** The symbol 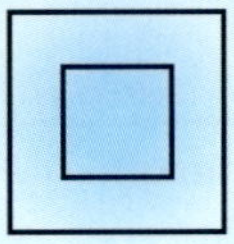 on an appliance means that it does not need

**A** a live wire

**B** a neutral wire

**C** an earth wire

**D** a fuse

**6** A fuse has a rating of 5 A.

Which current is most likely to cause the fuse wire to melt?

**A** 1 A

**B** 3 A

**C** 5 A

**D** 10 A

**7** When a direct current of 4 A passes in a bar heater, the wire is heated and glows red.

If an alternating current of 4 A passes in the same heater, the wire

**A** is cooled

**B** is heated but does not glow

**C** is heated and glows red

**D** is heated and melts

**8** The cost of energy from the electricity supply is about 7p for each

**A** amp

**B** joule

**C** kilowatt-hour

**D** volt

**9** In the National Grid, voltages are increased and decreased using a

**A** transformer

**B** transistor

**C** transmitter

**D** transpirer

**10** Most of the electricity in the UK is generated by burning

**A** fossil fuels

**B** hydroelectric fuels

**C** uranium

**D** renewable fuels

**11** A power station generates electricity by burning wood from fast-growing willow trees. This fuel

**A** does not produce any carbon dioxide

**B** does not leave any ash

**C** is non-renewable

**D** is renewable

**12** Double glazing is a form of insulation used

**A** in a loft

**B** in windows

**C** between the inner and outer walls of a house

**D** in floors

## Standard demand

**13** A current of 3 A passes in a lamp. The resistance of the filament is 12 Ω. The voltage across the filament is

**A** 4 V

**B** 9 V

**C** 15 V

**D** 36 V

**14** A diode

**A** only allows direct current to pass

**B** only allows alternating current to pass

**C** allows current to pass in either direction

**D** allows current to pass in one direction only

**15** Which component has this circuit symbol?

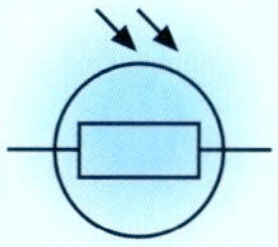

**A** a diode

**B** a fixed resistor

**C** a light-dependent resistor

**D** a thermistor

**16** Which circuit is suitable for measuring the resistance of a filament lamp?

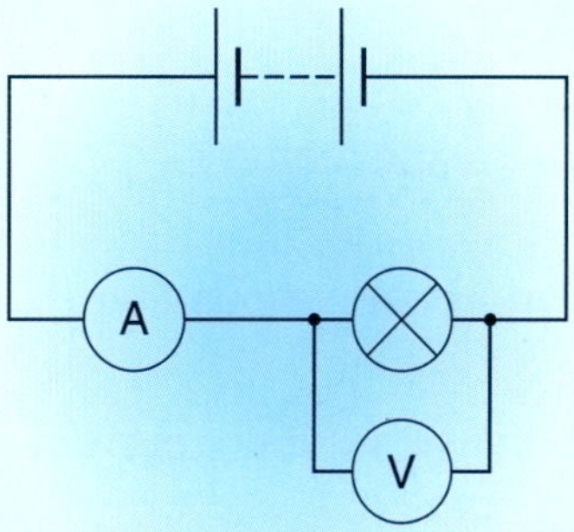


**A**

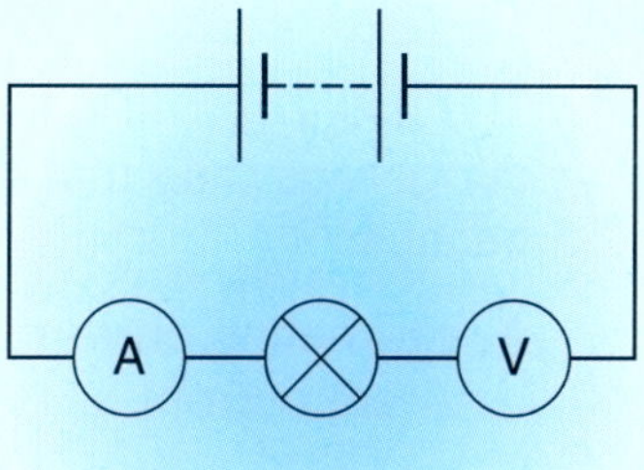


**B**

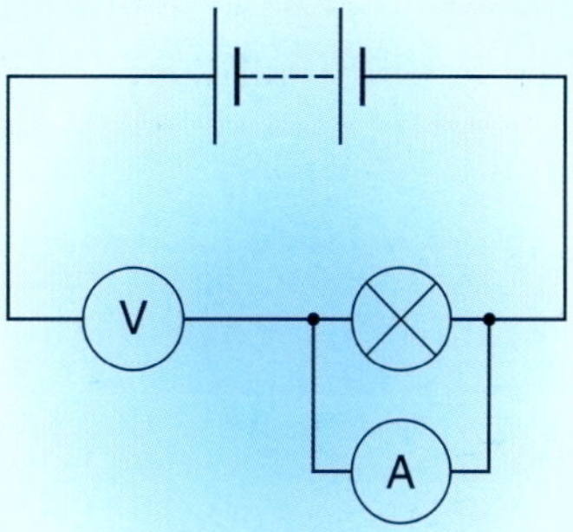


**C**

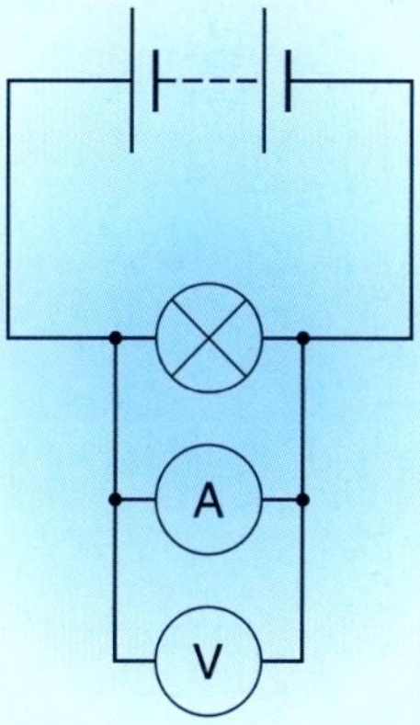


**D**

**17** Which is a correct description of the purpose of a fuse?

**A** to control the current

**B** to control the voltage

**C** to control the power

**D** to prevent fire

**18** The following are used to connect an appliance to the mains supply

**A** earth wire

**B** fuse

**C** live wire

**D** neutral wire

Which of these normally carries no current?

**19** When a current passes in a wire there is always some energy transfer from the electricity supply to

**A** heat

**B** light

**C** movement

**D** sound

**20** The 240 V mains a.c. supply can be more dangerous than that of a 9 V d.c. battery because

**A** alternating current is more dangerous than direct current

**B** the mains supply can cause a greater current than the battery can

**C** humans have a higher resistance to direct current than to alternating current

**D** humans have a lower resistance to direct current than to alternating current

**21** Which of these is a non-renewable source of energy?

**A** nuclear fuel

**B** tides

**C** waves

**D** wind

**22** The energy input to a wind generator is

**A** electricity

**B** gravitational potential energy

**C** kinetic energy

**D** sound

**23** Electricity is generated in a dynamo by

**A** a rotating electric field

**B** a stationary electric field

**C** a rotating magnetic field

**D** a stationary magnetic field

**24** An advantage of using low-energy light lamps instead of filament lamps is

**A** they are cheaper to buy

**B** they give out more light

**C** they give out more heat

**D** they are more efficient

# Module 6 – Waves, atoms and space

Modern living relies on being able to communicate using telephone, radio and television. These all use **waves**. The behaviour of waves and the uses of different types of wave is the first topic in this module. You will learn about which waves are safe to use and which need to be used with care to avoid unnecessary danger.

Beyond our planet and Solar System there is a vast Universe. New discoveries about what lies in outer **space** are being made every day. In the second topic in this module you will learn about the different types of objects and their orbits, as well as what holds the Universe together.

By studying the structure of the **atom**, the subject of the third topic in this module, you will learn about radioactivity. Radioactive materials have many uses that benefit humans, particularly in medicine, but they need to be used carefully.

Try these questions first.

1 Which can travel in a vacuum, light or sound?

2 The amplitude of a sound wave is increased. How does this affect the sound that is heard?

3 Light changes speed when it crosses a boundary. What is the name of this effect?

4 Which object is at the centre of the Solar System?

5 What does a planet orbit?

# 6:1 Wave measurements

Waves are caused by a disturbance travelling across the surface of the sea.

Light, sound waves and radio waves are all used for communicating. Unlike waves on a rope or the surface of the sea, there is nothing obviously wavy about them!

Studying waves that you can see helps you to understand the behaviour of those waves that you cannot see.

The photograph shows waves on the surface of the sea. On a calm day the surface of the sea is flat and undisturbed. On a rough day, the waves travel across the surface, making the gulls and the boat go up and down.

**Did you know?**
Heinrich Hertz discovered radio waves in 1888. He noticed that there was a small current in an electric circuit when he made a spark in the air nearby. He realised that something invisible had passed through the air.

## Measuring waves

If you fix one end of a long spring or rope, and move the other from side to side, a wave travels along the spring.

By moving your hand from side to side more rapidly, you will make more waves each second. You have increased the **frequency** of the wave.

The frequency of a wave is the number of waves produced each second. It is measured in hertz (Hz). One hertz means one wave per second.

**a If you move the end of the spring from side to side twice each second, what is the frequency of the waves produced?**

The diagram shows the movement in a wave travelling along a rope.

There are two quantities we can measure on this diagram:

- **Amplitude**. This is the greatest amount that any part of the rope moves from its normal position.
- **Wavelength**. This is the length of one complete wave – a crest and a trough.

The amplitude and wavelength of a wave are both measured in metres (m).

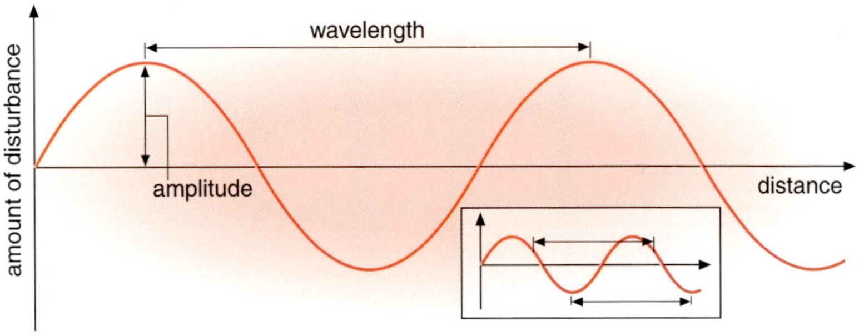


Take care! The amplitude of a water wave is measured from the undisturbed level of the surface.

**b** Look at the diagram of the wave. Which is greater, its amplitude or its wavelength?

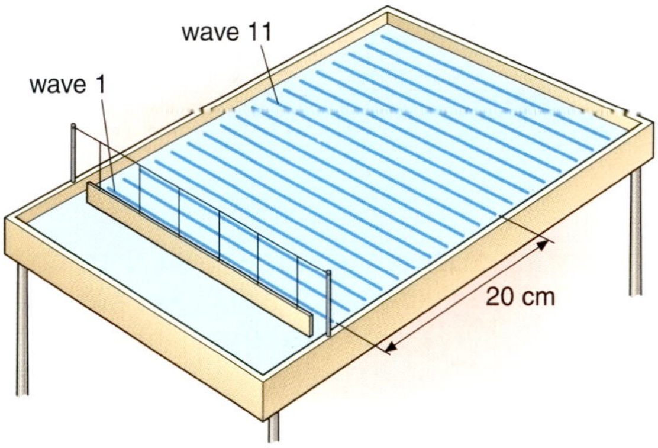


## Making waves

You can see waves on the surface of water using a ripple tank. A straight bar moves up and down, disturbing the surface of the water so that waves travel across it. The waves are projected onto the floor, or onto a screen. To find the wavelength, measure across several waves.

In the picture, there are 20 cm between wave 1 and wave 11. There are 10 waves in this distance. So:

Wavelength = 20 cm ÷ 10 = 2 cm.

## Different types of wave

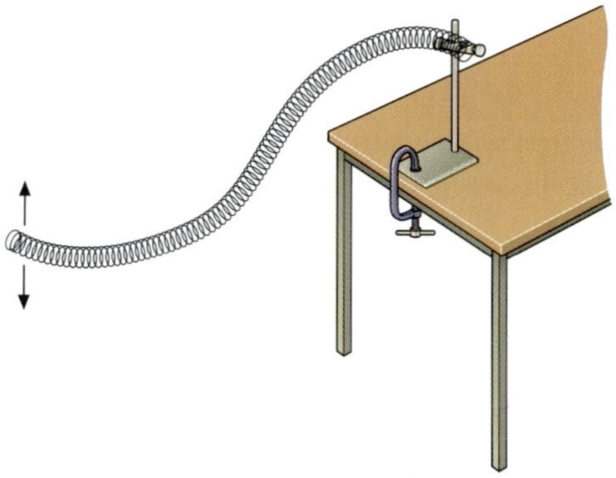

When you hold one end of a spring and move your hand from side to side, the wave that travels along the spring is a **transverse** wave.

In a transverse wave, the **vibrations** are at right angles to the direction of wave travel. Light and radio waves are examples of transverse waves.

## Making longitudinal waves

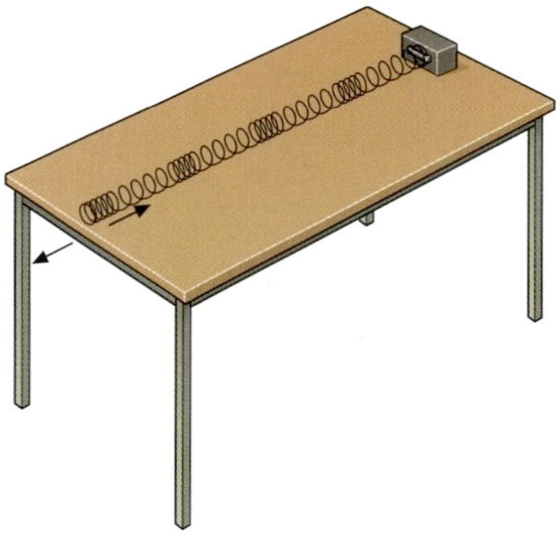

There is another way of making a wave on a stretched spring. If you push the end back and forth, along the length of the spring, the wave looks different from a transverse wave. A region of squashed-up spring travels along the length, followed by a region of stretched-out spring.

This is an example of a **longitudinal** wave.

In a longitudinal wave, the vibrations are along the direction in which the waves themselves are travelling. Sound travels as a longitudinal wave.

## Questions

**1** Copy the table. Fill in the correct units in the second column. Choose from hertz (Hz) and metres (m).

| Quantity | Unit |
|---|---|
| amplitude | |
| wavelength | |
| frequency | |

**2** Look at the diagram of a wave.

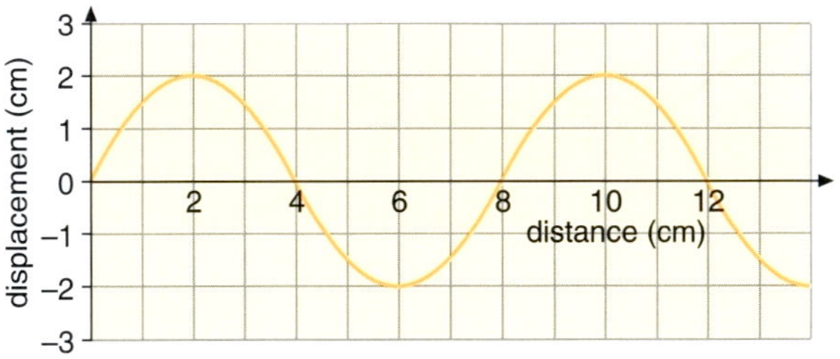


**a** What is its amplitude?

**b** What is its wavelength?

**3** Jo is making waves travel along a stretched length of rope. She moves her hand from side to side 30 times in 10 s.

**a** How many waves does she make in 1 s?

**b** What is the frequency of the waves?

### Key ideas

- *Amplitude, frequency and wavelength are important measurements that apply to all waves.*
- *Waves can be described as transverse or longitudinal.*

# 6:2 A family of waves

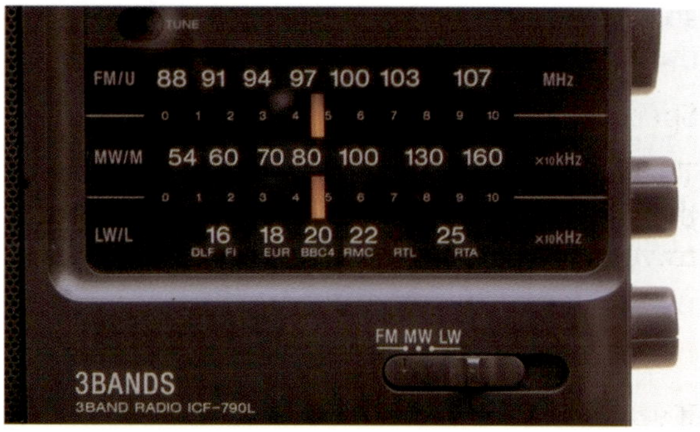


When you are listening to the radio, you can tune it to different stations. When you turn the dial, you are changing the frequency of the radio waves you are tuning to. On the FM scale, the frequencies range from about 88 MHz to about 106 MHz. They make up a **spectrum** of frequencies from which you can select the frequency of the station you want to hear.

**a What frequency does your favourite radio station broadcast on?**

You can make a spectrum of light by passing white light through a prism. Light travels as a wave. The light is split up into different colours, from red to violet. The waves of different colours have different frequencies and wavelengths.

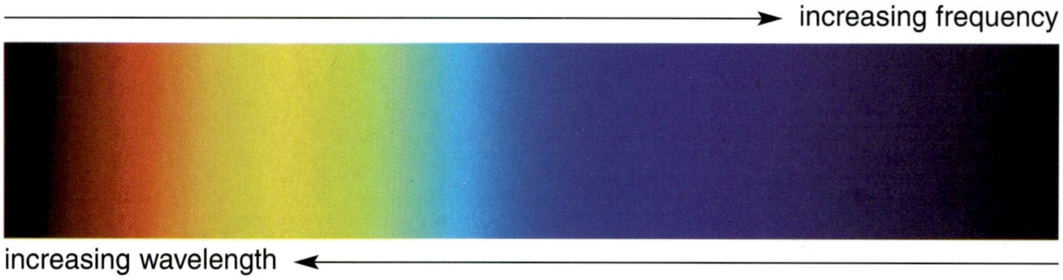


**b Which colour of light has the longest wavelength? Which colour has the shortest wavelength?**

> **Did you know?**
> Isaac Newton was the first person to realise that all the colours of the rainbow are contained within white light. Prior to this, people thought that coloured objects added colour to light, like adding a dye to paint.

## More waves

**Light** and **radio waves** are just two types of **electromagnetic wave**. There are several others, and together they make up the **electromagnetic spectrum**. The spectrum shows all types of electromagnetic radiation, arranged in order of their wavelengths and frequencies.

You will have heard of each of these types of radiation, but you may not have realised that they all belong to one big family of electromagnetic waves.

highest frequency — lowest frequency

| gamma rays and X-rays | ultraviolet radiation | visible light | infrared radiation | microwaves | radio waves |
|---|---|---|---|---|---|

shortest wavelength — longest wavelength

**c Which type of electromagnetic wave has the longest wavelength? Which have the shortest wavelength?**

## Top speed

Electromagnetic waves are transverse waves. Although we cannot see their vibrations, these are at right angles to the direction of wave travel. They also travel very fast. In empty space there is nothing to slow them down, and they all travel at the same speed:

Speed of electromagnetic waves in space = 300 000 000 m/s

That's 300 million metres per second, or over seven times round the Earth in one second. Most scientists think that nothing can travel faster than this. It is the speed limit of the Universe.

## Getting warm

If you lie in the sun, the electromagnetic radiation that falls on your skin includes:

- light;
- **infrared** radiation;
- **ultraviolet** radiation.

These are all absorbed by your skin and make you feel warm. The energy carried by the waves is transferred to your body.

For all electromagnetic waves:

- the shorter the wavelength of the radiation, the higher its energy;
- the higher the energy, the more danger the wave poses to body tissue.

Radio waves are absorbed by a metal aerial. The waves make electric charge move to and fro in the aerial. This is an alternating current (because it changes direction), and it has the same frequency as the radio waves.

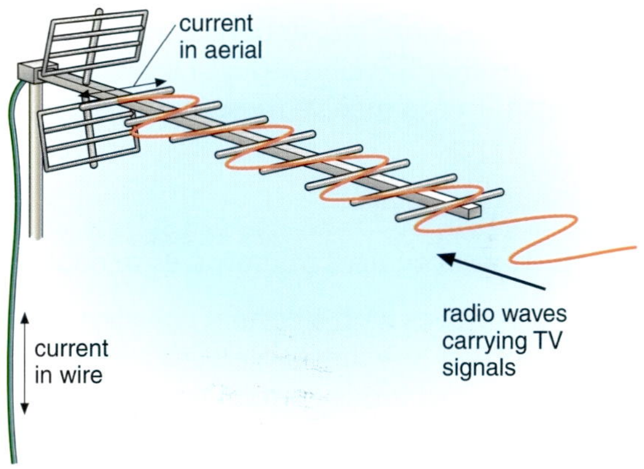


**Television signals are carried by high frequency radio waves. They cause an alternating current in the aerial, and this passes in a wire leading to the TV set.**

## Questions

1 Put the following types of electromagnetic radiation in order, starting with the ones with the longest wavelength.

| | |
|---|---|
| light | radio waves |
| infrared radiation | ultraviolet radiation |
| gamma rays | microwaves |

2 Which have a higher frequency, gamma rays or ultraviolet radiation?

3 Which electromagnetic waves have a shorter wavelength than that of light?

4 a How fast do radio waves travel in empty space?

b If a spacecraft sent some radio waves and some microwaves through space to Earth, which would get there first?

## Key ideas

- *Electromagnetic waves are transverse waves. They all travel through space at the same speed.*
- *They form a continuous spectrum:*

*highest frequency/shortest wavelength*

*gamma rays*

*X-rays*

*ultraviolet radiation*

*light*

*infrared radiation*

*microwaves*

*radio waves*

*lowest frequency/longest wavelength*

# 6:3 The longer waves

Guglielmo Marconi was the first person to send messages using radio waves. In 1901 he sent the first radio message across the Atlantic. Many people thought that his signal would travel straight out into space, because the Earth is curved.

The radio waves that he was using had a long wavelength. These radio waves follow the Earth's curvature and so they can be sent between distant points on the Earth's surface.

Some radio stations use long wavelength radio waves to broadcast over large areas. On holiday, you might be able to pick up a station from your home country even if you are thousands of kilometres away.

FM radio and television stations use much shorter wavelength radio waves that travel in straight lines, so you need to be close to a transmitter to receive these stations.

Marconi sent the first transatlantic radio messages. One of his signals helped catch Dr Crippen, a murderer who had escaped from England on a ship. Detectives were ready to arrest him when he arrived in New York.

## Using microwaves

A **microwave** oven uses microwaves to cook and re-heat food.

- Microwaves with a certain wavelength and frequency are absorbed by water molecules in food.
- Since most foods contain water, this is an efficient way of transferring energy to food.
- China plates and plastic containers do not absorb the microwaves.

Microwaves are short-wavelength radio waves and they are used for communicating. Microwaves are used to send national television broadcasts from the Telecom tower to local transmitters.

**Did you know?**
Long wavelength waves are used to transmit information to shipping because they can be received a long way from the transmitter.

Mobile telephones use microwaves to transmit messages to a base station. More microwaves are used to relay the message to another more distant base station, close to the receiving phone. Then it is sent to the receiving phone.

Microwaves can pass easily through the atmosphere. They are not reflected (unlike some radio waves). This means that they are useful for sending signals to satellites, which orbit high above the atmosphere. Satellites amplify the signals and use microwaves to send them back down to Earth.

**a Why are scientists concerned about young people using mobile telephones?**

These 'dishes' are microwave aerials. Masts like these are usually found on hilltops. They relay the signals to other dishes many kilometres away.

This satellite is in orbit far above the Earth. It receives television programmes sent as microwave signals from Earth, and broadcasts them back down to Earth.

## Infrared radiation

All objects give out **infrared radiation**. The hotter the object, the more radiation it gives out. The glowing element in an electric toaster is very hot – it gives out intense infrared radiation. This is absorbed by the bread that is being toasted, so that the bread becomes hot and burns gently.

Other uses of infrared radiation include:

- Night vision. Infrared cameras can detect people and other animals because they emit more infrared radiation than their surroundings.
- Remote controls. These send messages in code to your television or hi-fi system.
- Security systems. Infrared detectors can switch on lights and sound alarms when they sense the radiation from a person's body.
- Muscle treatment. The heating effect of infrared radiation can help damaged muscles to heal quickly.

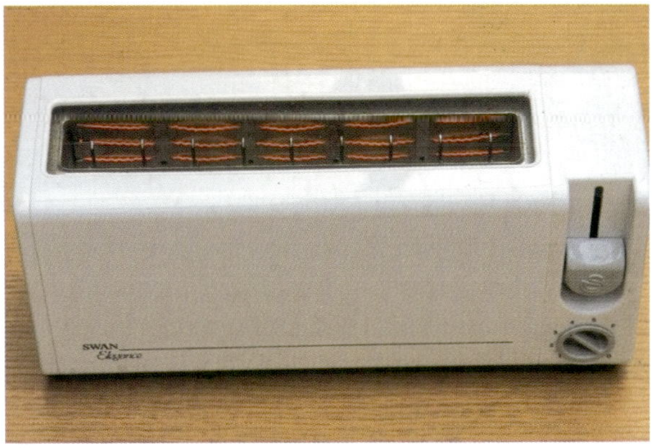

The element in this toaster is hot – about 700 °C. It emits two types of electromagnetic radiation: light and infrared.

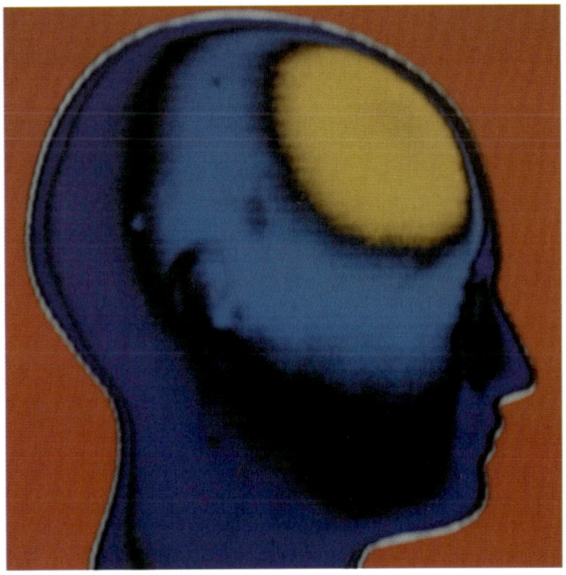

On this infrared photograph, the brain tumour shows up because it is warmer than the surrounding tissue.

**b** Some cookers are fitted with electric grills. You can see them glowing as they cook the food. What type of electromagnetic radiation can you see? What other type of electromagnetic radiation does a grill emit?

## Questions

1 What properties of microwaves make them suitable for satellite transmissions?

2 a Why was Marconi able to send a radio transmission across the Atlantic Ocean, despite the Earth's curvature?

b What type of wave would we use now to send a radio transmission across the Atlantic, and how would it travel?

3 A national radio station is broadcast on two different wavelengths: 3 m and 1500 m.

Which of these can be detected hundreds of miles from the transmitter? Explain why.

### Key ideas

- Radio waves and microwaves are used for communications systems.
- Microwaves and infrared radiation are used in cooking.
- Remote controls use infrared radiation.

# 6:4 Shorter than light

The ozone layer in the atmosphere protects us from the Sun's most harmful radiation, but there is a man-made hole in it. You need to take care to avoid being exposed to too much ultraviolet radiation. Some people use sun beds to give themselves a tan, but it isn't very sensible to increase the amount of **ultraviolet radiation** that you are exposed to.

Suntan lotion contains chemicals that absorb ultraviolet radiation.

## Short wavelengths

Ultraviolet radiation is like light, but with shorter wavelengths. It is invisible to the human eye (though some other creatures can see it).

**a Explain why a special lamp is needed to show up the security paint (shown in the photograph).**

To read this security marking, shine ultraviolet radiation on it. The paint absorbs the ultraviolet radiation and re-emits it as visible light.

## Even shorter

**X-rays** and **gamma rays** have the shortest wavelength of all the electromagnetic waves. This makes them potentially dangerous. They are in the same part of the spectrum. There is no difference in the waves themselves, but they are produced in different ways.

- X-rays come from X-ray tubes that can be turned on and off.
- Gamma rays are given out by an unstable nucleus in radioactive decay. This cannot be turned off.

Both types of radiation are very penetrating. They pass easily through flesh, but not so readily through bone or metal.

This makes X-rays useful for taking a shadow photograph to detect a bone fracture and for examining luggage to detect hidden metal objects.

**b Explain how X-rays would be useful if a small child swallowed a coin.**

X-rays produce a shadow picture, revealing the bones of living creatures.

## Damaging cells

The more radiation that people are exposed to, the more harm it may do. The paler the skin, the deeper ultraviolet radiation can penetrate, and the greater the chance that a cancer may be caused. Cancers are caused when high-energy waves cause a particular mutation to a gene. This makes cells reproduce more rapidly than normal.

Ultraviolet radiation can be particularly damaging to the eye, so you should never look directly at the Sun.

- Low doses of ultraviolet, X-radiation and gamma radiation can damage normal cells so that they become cancerous.
- Higher doses of these types of radiation can kill cells.
- Over-exposure to infrared radiation causes skin burns.
- Microwaves can damage body tissue by heating it.

Sometimes we want to kill cells. Some foods are exposed to gamma radiation to kill bacteria and extend their shelf life. The food can then be safely given to seriously ill people or taken on space flights. In the same way, surgical instruments are sterilised using gamma radiation.

**c** Why is it important to sterilise surgical instruments before use?

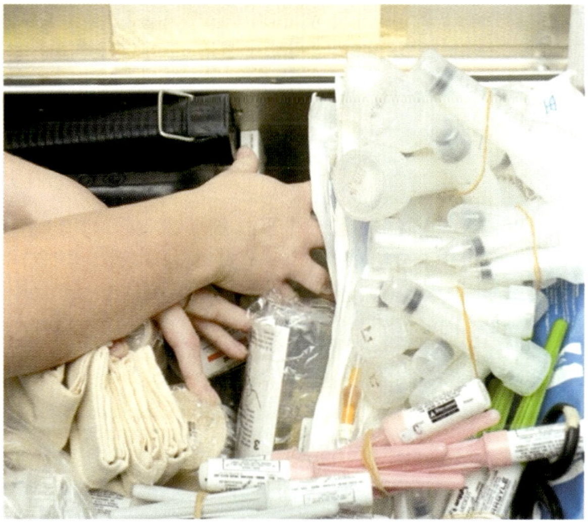

Gamma radiation has been used to sterilise these plastic syringes inside their packaging.

**Did you know?**

Irradiated foods that are sold in supermarkets have to be specially labelled so that shoppers know what they are buying.

## Treating cancer

It may seem strange that radiation that can cause cancer can also be used to cure it. X-rays and gamma rays can kill any type of body cell if the dose is high enough.

- Beams of X-rays or gamma rays are directed at a tumour.
- Several beams are used that overlap at the tumour.
- In this way the tumour receives a much higher dose of radiation than the surrounding tissue.

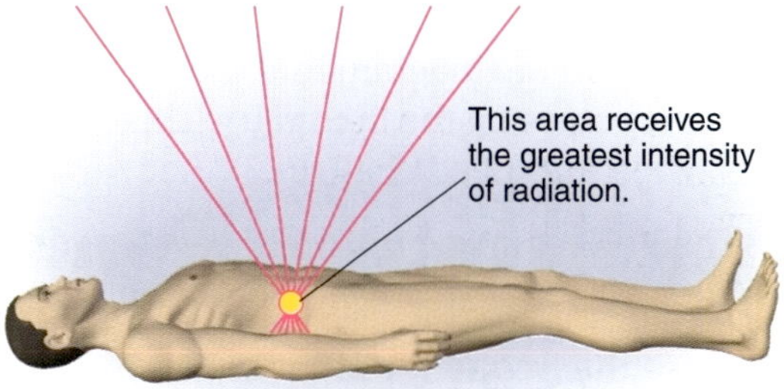


## Questions

1 Copy the sentences below and use words from the list to fill the gaps.

killed absorbed cancerous

When radiation penetrates living tissue, it may be __________ or it may pass straight through.

A low dose may cause cells to become __________, while cells may be __________ by a higher dose.

2 Why are the short wavelength radiations potentially more hazardous than light?

3 When taking an X-ray of a patient's teeth, a dentist stands outside the room, or behind a protective metal screen.

a Why might the dentist leave the room?

b Why should the protective screen be made of metal?

4 Describe what happens when

a infrared radiation lands on your skin

b ultraviolet radiation lands on your skin.

### Key ideas

- *Radiation may be absorbed by living tissue or it may pass through.*
- *When radiation is absorbed by cells, they may be killed by it; lower doses may cause the cells to become cancerous.*

# 6:5 Changing direction

Seeing involves the eyes and the brain. The eyes sense light and the brain puts all the information together to make a picture. To put things in the right place the brain assumes that light travels in straight lines.

Sometimes it does, but when it does not, the brain gets things in the wrong place.

There are lots of everyday examples of this. Every time that you look into a mirror you see an image behind the mirror. But it is not there. You know that swimming pools and ponds always look shallower than they really are.

These effects are caused by light changing direction, either when it is reflected or when it passes through different materials.

**This person is looking at an image that is not real.**

> **Did you know?**
> When viewed through water, objects seem to be closer by about one quarter the depth of the water.

## Slowing down

Waves *slow down* when they pass from air into glass or water. When the speed of a wave changes it is called **refraction**. All waves can be refracted. The diagram shows what happens to waves of light going into the glass of a window.

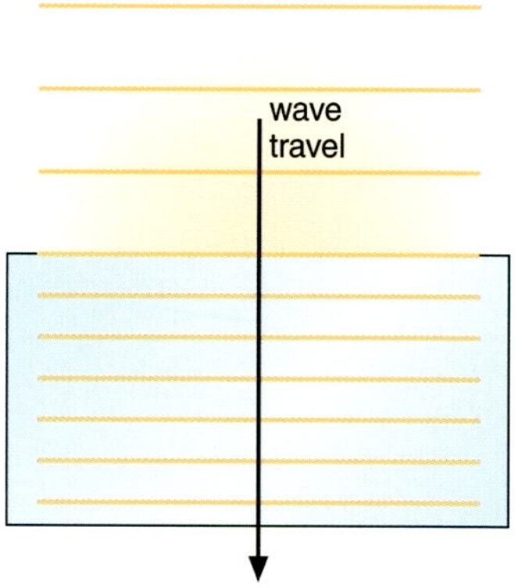


You can see from the diagram that the waves look to be closer together after crossing the boundary. This is because the wavelength changes when the speed changes.

- When waves *speed up*, their wavelength *increases*.
- When waves *slow down*, their wavelength *decreases*.

In each case the frequency of the waves is not affected, the number of waves leaving the glass each second is the same as the number entering.

**a Which two wave measurements always change when a wave is refracted?**

## Changing direction

When a wave is refracted, the change in speed may also cause a change in direction. This happens when the direction of wave travel is not along a **normal line** – a line drawn at right angles to the surface.

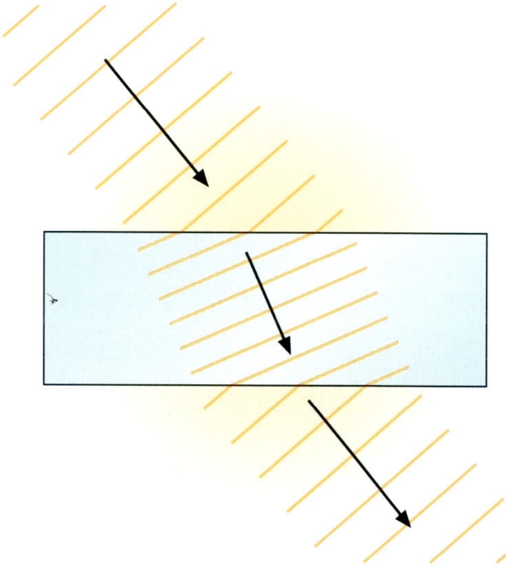

The diagram shows the change in direction when waves are slowed down and when they speed up again.

Light leaves a window in the same direction that it enters. This is because the sides of the window are parallel. As the light leaves the window, the change in direction is opposite to the change as it goes in.

When you look through a window, you do not notice the effects of refraction. With other shapes of glass there is a noticeable change in direction.

**b** **Draw the appearance of the light waves as they pass through the prism.**

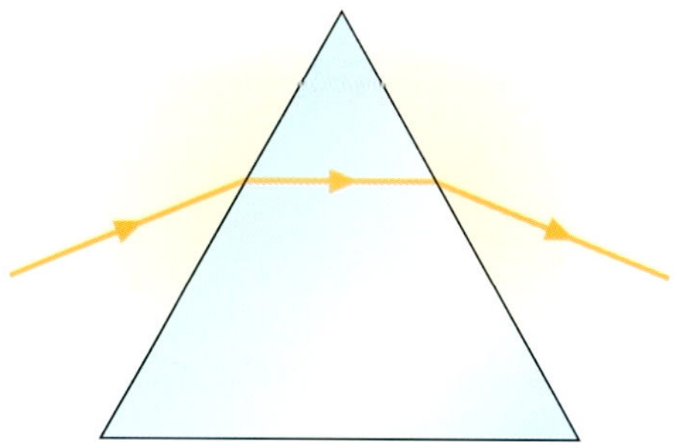

## Using optical fibres

Both light and infrared radiation are used to send information along **optical fibres**. These are the advantages of using optical fibres.

- Light can travel round corners in an optical fibre, so unlike microwave transmissions in air, it is not restricted to travelling in a straight line.
- Light passes along an optical fibre with very little energy loss, so unlike signals in copper cables, the signal does not need frequent amplification.

The diagram shows a **digital signal** travelling along an optical fibre. Digital signals are replacing **analogue signals** in communications.

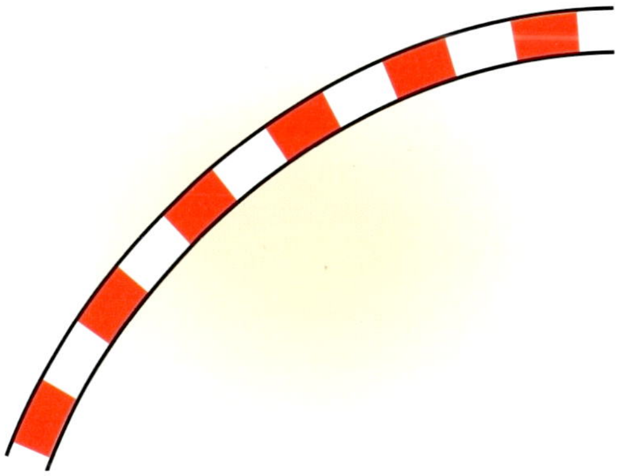

- A digital signal can only have certain values, usually 0 or 1.
- An analogue signal carries information by continuously changing its amplitude or frequency.

**c** **An AM radio wave has a variable amplitude. What varies in an FM radio wave?**

Noise distorts all signals during transmission. It cannot be removed from analogue signals, but digital signals can be restored to their original condition, giving clearer sound and pictures.

An analogue signal.



A digital signal.

## Questions

**1** **Copy the sentences and complete them using words from the list.**

**direction**  **frequency**  **speed**  **wavelength**

**Refraction describes a change in the ________ of a wave. When a wave is refracted, its ________ also changes and its ________ may change. The ________ of the wave never changes.**

**2** **Copy and complete the diagrams to show light travelling through glass.**

**3** **What is the advantage of using optical fibres to send information**

**a** **in place of microwaves**

**b** **in place of copper cables?**

### Key ideas

- *Refraction describes the change in speed of a wave.*
- *Information can be sent along optical fibres with little energy loss.*
- *Both analogue and digital signals are used in communications.*

# 6:6 Sound and ultrasound

When you speak to a group of friends the air between your mouth and their ears is alternately squashed up and stretched out. These compressions and stretches travel through the air, making the air particles vibrate. Sound is a longitudinal wave, as the vibrations are along the direction of wave travel.

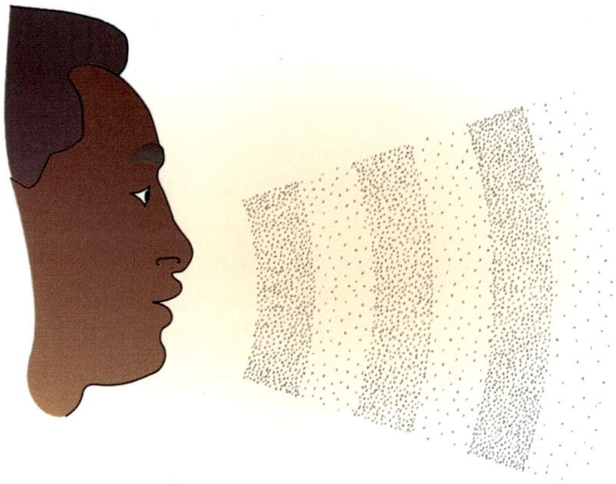

The squashes and stretches in a sound wave.

Our ears cannot detect all compression waves that travel through the air – only those with the right frequency.

For a young person with good hearing, sounds with frequencies in the **range** 20 Hz to 20 000 Hz are heard as sound. As a person ages and parts of the ear wear, sounds at the upper end of this range can no longer be heard.

Our ears are more sensitive to some frequencies than others. The graph shows how the sensitivity of a young person's hearing compares to that of an older person.

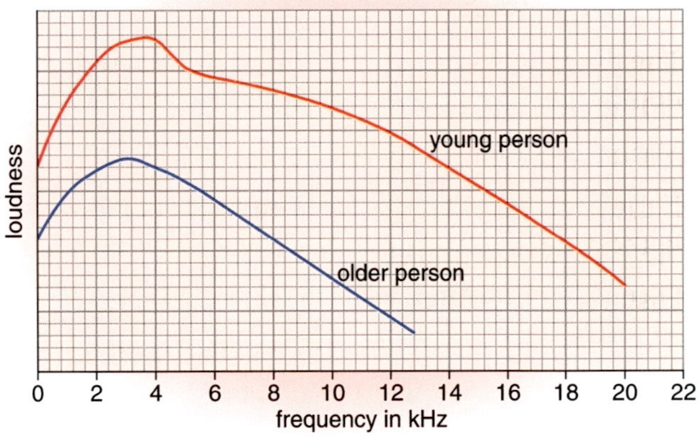


**a What is the highest frequency sound that the older person can hear?**

## Beyond hearing

Bats produce high-pitched sounds, too high for us to hear. They listen for the echoes from their surroundings, and this helps them to 'see' their surroundings and find food. Any sound that is too high for humans to hear is called **ultrasound**. The frequency of ultrasound is higher than the upper limit of our hearing range, so ultrasound has frequencies greater than 20 000 Hz (20 kHz).

Many creatures, such as this bat, produce ultrasound.

**Did you know?**
Elephants and dolphins communicate using infrasound, sound waves that have a frequency below 20 Hz.

**b Which of the following frequencies could be the frequency of an ultrasound wave?**

**150 Hz, 15 kHz, 300 Hz, 30 kHz**

## Using ultrasound

Ultrasound has several uses. It is used to produce an image (a scan) of a fetus in the uterus. The operator moves a probe over the woman's uterus. The probe gives off ultrasound and detects reflections. These are then used to build up a picture on a screen.

Ultrasound is better than X-rays for this purpose; there is a slight risk to mother and baby from X-rays, but there is no evidence of any risk from ultrasound.

The same technique can be used for other medical purposes, such as looking for gallstones in the gall bladder or finding the precise position of a tumour in the brain.

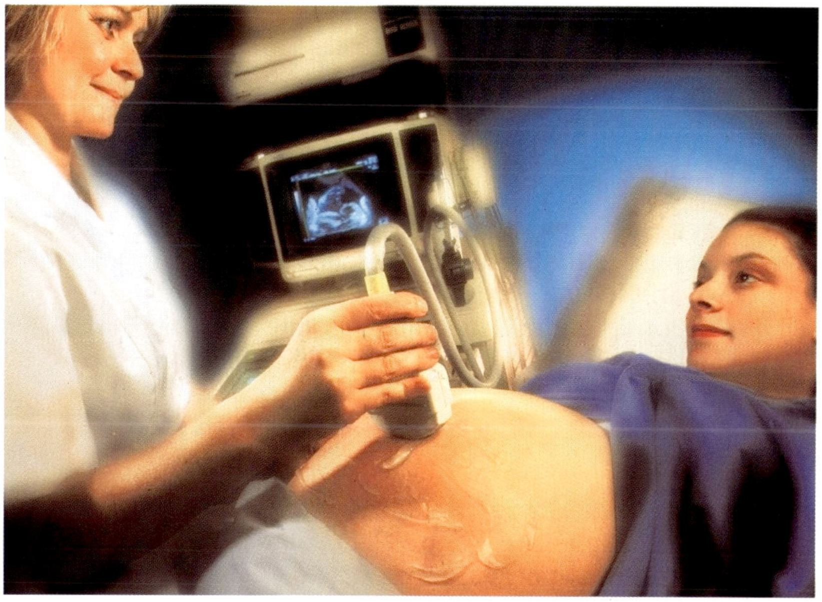

Pre-natal scanning. Ultrasound waves reflect off the fetus in the uterus and are detected, to produce an image on the screen.

## Echo sounding

In a similar way, ultrasound is used in **echo sounding**. A distance is measured by measuring the time delay between when a pulse of ultrasound is sent out and when the echo is received. Echo sounding is used:

- to detect cracks and flaws in railway lines;
- to measure the depth of the sea or a lake;
- to find shoals of fish or the remains of ships that have sunk.

The diagram shows a ship using ultrasound to measure the depth of the sea.

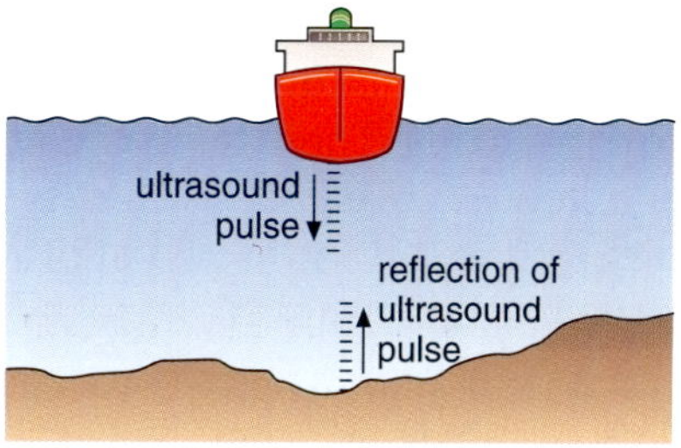


Measuring the depth of the sea.

Sound travels in water at a speed of 1500 m/s.

If the time between sending out a pulse of ultrasound and detecting the echo is 0.8 s:

Distance travelled by pulse = speed × time

= 1500 m/s × 0.8 s

= 1200 m.

The depth of the sea is just half this distance, 600 m, because the pulse has to travel there and back again.

**c How will the ultrasound operator on the ship tell when a shoal of fish swims under the ship?**

## Questions

**1** What is the approximate range of frequencies of sound that can be detected by a young person with good hearing?

**2** Explain what is meant by ultrasound.

**3** Ultrasound waves are used in pre-natal scanning.

**a** What does 'pre-natal' mean?

**b** Why is ultrasound scanning safer than using X-rays?

**4** Sound travels in water at 1500 m/s.

A ship sends out a pulse of ultrasound and detects the echo from the seabed 1.2 s later.

**a** How far has the ultrasound travelled?

**b** How deep is the sea?

### Key ideas

- *Sound travels as a longitudinal wave.*
- *Ultrasound waves are sound waves with frequencies higher than the limit of human hearing.*
- *Ultrasound can be used in medicine to produce images of the inside of the body, and for echo sounding.*

# 6:7 Around the Sun

No-one has ever seen the **Solar System** looking like this. You would have to travel far out into space to be able to look down and see the **planets** in their **orbits**. It is hard to tell, but the orbits of the planets are not quite circular. Their shape is called an **ellipse**, a squashed circle, with the Sun close to the centre. Pluto and Mercury have the most elliptical orbits, the orbits of the other planets are close to circles.

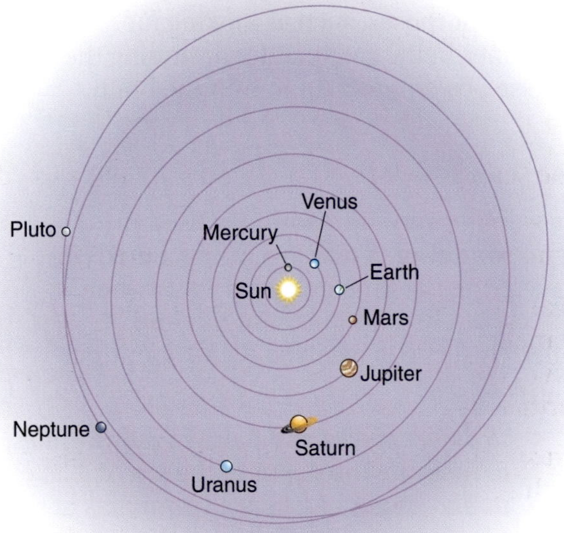


The closer a planet is to the Sun, the faster it travels and the quicker it completes one orbit. Pluto takes the longest time for one orbit – almost 250 years!

**a Which planet is closest to the Sun? Is Pluto always furthest from the Sun?**

## Held by gravitational force

If you want to whirl a rubber bung around in a circle, you have to keep pulling on the string. There has to be an inward force to make any object follow a curved path. There is no piece of string joining the Earth to the Sun. So what provides the pulling force that keeps the Earth in its orbit?

The answer is **gravitational attraction** – forces that act between all objects with any mass. We do not notice these forces between everyday objects because their masses are small. But as the masses get bigger, so do the attractive forces.

These cars are attracting each other, but the forces are not big enough to overcome the friction forces that stop them moving together.

The Sun is a very massive object. This gives it a strong **gravitational pull**. The gravitational pull of the Sun on the Earth keeps it going round in its orbit.

The diagram shows the gravitational forces between the Sun and the Earth.

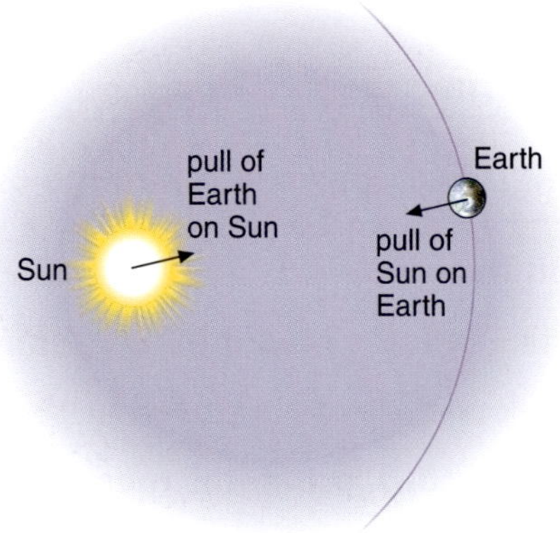


This diagram reminds you that:

- The Sun pulls the Earth and the Earth pulls the Sun.
- These forces are equal in size and opposite in direction.

**b What type of objects can pull with gravitational forces?**

**What type of objects are pulled by gravitational forces?**

## Satellites

A **satellite** is an object that orbits another object. There are many artificial satellites in orbit around the Earth. The **Moon** is often described as the Earth's natural satellite because it was not put into orbit by people.

Other planets also have moons – Jupiter has sixteen of them.

**Did you know?**
Pluto and its moon, Charon, are very similar in size and mass. They move around like a spinning dumbbell.

The forces that keep satellites and moons in orbit around planets are just like the force that keeps the Earth in orbit round the Sun. They are gravitational attractive forces. The diagram shows the gravitational forces between the Earth and the Moon, and between the Earth and an artificial satellite.

## How strong is the pull?

The size of the force that pulls a moon or a satellite depends on:

- the mass of the satellite;
- the strength of the planet's **gravitational field**.

Like the magnetic field of a magnet, the gravitational field of a planet extends into the space around it. It affects all things that have mass and, also like a magnet, it becomes weaker as you go further away from a planet.

**Gravitational field strength**, *g*, is the pulling force acting for each kilogram of material.

The table shows the gravitational field strength at the surface of some planets and our Moon.

| | Earth | Moon | Mars | Jupiter | Neptune |
|---|---|---|---|---|---|
| **Gravitational field strength, *g*** | 9.8 N/kg | 1.6 N/kg | 3.7 N/kg | 23 N/kg | 11 N/kg |

Now you can understand why an astronaut finds it easy to walk on the Moon. Weight is just the size of the gravitational pull.

On Earth, an astronaut and his space suit could have a mass of 90 kg.

His weight = 90 kg × 9.8 N/kg = 882 N.

**c How much would the astronaut weigh on the Moon?**

## Questions

1 Draw a diagram that shows the gravitational forces between the Sun and Mercury, the planet closest to the Sun. Write a short description of each force.

2 Copy the sentences. Use words from the list to complete them.

moons satellites Solar System Sun

Everything that goes around the Sun is in the ____________. Planets are orbited by ____________ and ____________. Planets themselves orbit the ____________.

3 a What would a 90 kg astronaut weigh if he could stand on the surface of Jupiter?

b What difficulties might this cause?

### Key ideas

- *Moons and satellites orbit planets.*
- *Planets orbit the Sun.*
- *These orbits are due to gravitational attractive forces.*

# 6:8 A wider view

Planets are not the only objects that orbit the Sun. Asteroids are lumps of rock, most of which lie in the asteroid belt between Mars and Jupiter. At least one of these even has its own satellite.

There are an unknown number of **comets**. We rarely see a comet, but when we do it is a spectacular sight.

A glowing comet is a spectacular sight in the night sky.

## Icy wanderers

A comet is a frozen ball of ices and dust. Comets orbit the Sun, but unlike the planets:

- their orbits are highly elliptical;
- they can orbit in any plane and any direction.

The shape of a comet's orbit means that the Sun's gravitational pull on it is always changing. This pull changes in size and in direction as the comet moves.

- As a comet approaches the Sun, the gravitational pull increases.
- As a comet moves further away from the Sun, the gravitational pull decreases.

**a At what point on the orbit of a comet is the Sun's gravitational pull:**

**i at its strongest**
**ii at its weakest?**

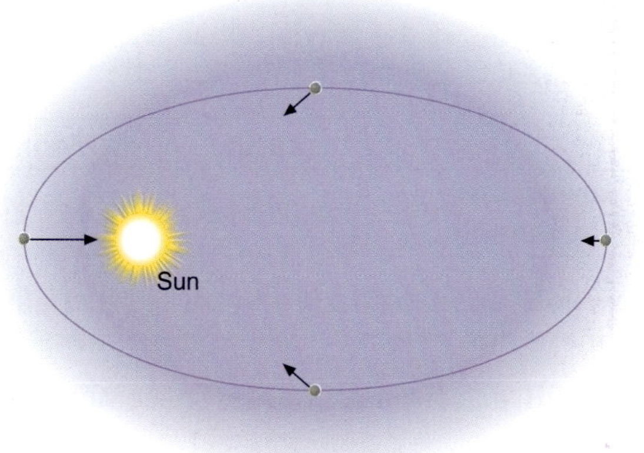


The changes in the Sun's pull on a comet.

**Did you know?**
A comet can be seen when it is close to the Sun. Energy from the Sun warms the comet and vaporises some of the dust and ices. These then glow as they reflect light from the Sun.

## Stars and galaxies

The Sun is a **star** – a fairly average star. Beyond the Solar System, there are billions more.

It can be difficult to see the stars in the night sky. It may be cloudy, and streetlights may fill the sky with light. But if you can find a dark place on a clear night, you'll see that the sky is filled with hundreds of stars.

The Sun belongs to a vast group of stars called the **Milky Way**. You may be able to see the Milky Way as a broad, cloudy band of stars, stretching across the night sky. This cluster of billions of stars is our **galaxy**. A galaxy is a collection of stars held together by gravitational forces.

Different galaxies have different shapes. The Milky Way is a spiral galaxy, with our Sun and Solar System on one of its arms, close to the edge. The whole galaxy whirls around at a tremendous speed.

**b What holds the stars together in a galaxy?**

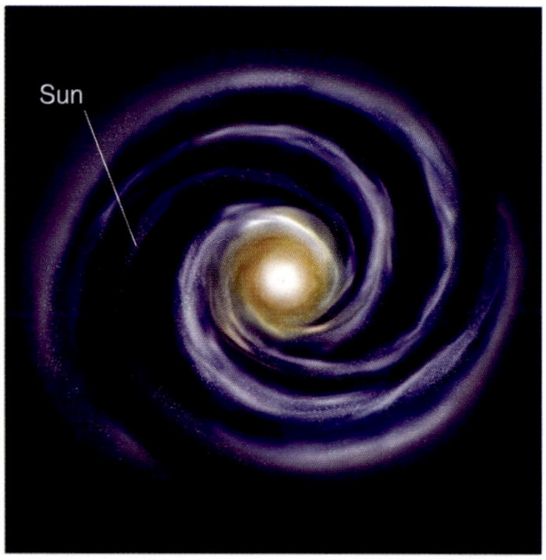


The Milky Way galaxy.

## More stars, more galaxies

With binoculars or a telescope, you can see many more stars in the night sky. These are all part of our own galaxy. But far beyond our galaxy there are other galaxies, and you may be able to see some of them if you know where to look.

Astronomers have seen and photographed many more galaxies. Some are spirals, like the Milky Way; others have more irregular shapes. They believe there may be as many as 100 000 million galaxies in the **Universe**.

The Universe is the whole of space. It contains all the matter and energy that exists.

Stars in one galaxy: 100 000 000 000

Galaxies in Universe: 100 000 000 000

So stars in Universe: 10 000 000 000 000 000 000 000

That's a lot of stars!

The bigger the telescope, the more light it can collect from distant stars.

**c** Why is it easier to see stars at night when you are out in the countryside than when you are in a town?

## Questions

**1** State two differences between the orbit of a comet and the orbit of a planet.

**2** Copy and complete these definitions. Choose words from the list to fill the gaps.

**the Universe** **a galaxy** **the Milky Way**

- **a** A group of millions of stars is called ________.
- **b** Our galaxy is called ________.
- **c** All the matter and energy that exists makes up ________.

**3** Imagine that you had an ultra-fast spaceship. Describe what you would see if you set off into space, travelling right across the Universe.

**4** Which contains most stars: a galaxy, the Solar System, the Universe?

### Key ideas

- *Comets move in elliptical orbits around the Sun. They can orbit in any plane and any direction.*
- *The Sun is just one of billions of stars which make up our galaxy, the Milky Way.*
- *There are millions of galaxies in the Universe.*

# 6:9 The life cycle of a star

Astronomers believe that our Sun is a fairly ordinary star, roughly half way through its life. But how did it begin? And how and why will it end?

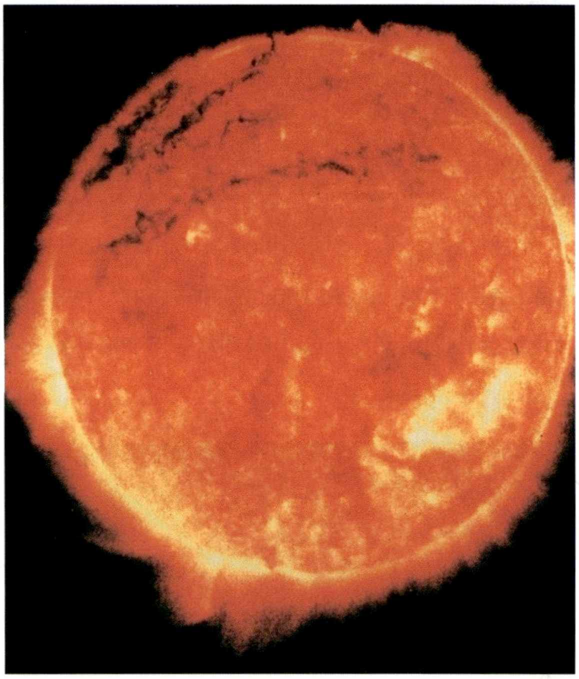

Space telescopes show that the surface of the Sun is patterned with hotter and cooler regions.

## Birth of a star

When astronomers use powerful telescopes to look at distant galaxies, they can see giant dust clouds where new stars are forming. This gives them an idea about how the Sun formed.

The Sun formed in a giant, swirling cloud of dust and gas called a **nebula**. In some parts of the cloud the gravitational attraction between the particles caused them to become closer and closer together.

- As the nebula contracted, it heated up.
- Eventually the temperature became great enough for particles to join together in a **nuclear fusion** reaction.
- Nuclear fusion reactions are the source of the Sun's energy.

New stars are forming inside this nebula.

**a What happens to the particles involved in a fusion reaction?**

## Billions of years

The Sun is in its **main sequence**. In its core protons are fusing together to form helium nuclei, releasing energy at a constant rate. This has lasted for about five billion years, and will probably last for another five billion years.

When there are no longer enough protons to keep the reaction going:

- The Sun will expand and cool, absorbing the planets as it becomes a **red giant**.
- Gravitational forces will then cause it to collapse to a small, very dense bright star – a **white dwarf**.
- As the white dwarf cools, it will dim to an invisible **black dwarf**.

**Did you know?**
The temperature in a nebula has to reach 15 million °C before fusion starts.

Contraction causes heating and expansion causes cooling. This is why the Sun will cool when it expands to a red giant. A white dwarf gives out light because it is very hot following the contraction, about 3000 °C. There are no fusion reactions going on to keep it hot and so it gradually fades as it cools.

As a white dwarf cools, its colour fades to orange and then red before it stops giving out any light at all.

**b What is the likely lifespan of the Sun, in billions of years?**

**c Which is hotter: a red giant or a white dwarf?**

## Larger stars

A star with more mass than the Sun has a different history. After becoming a red giant, it collapses inwards and then explodes outwards – a **supernova**. Gas and dust fly out into space. It is likely that our Solar System formed from the remains of a supernova that had exploded.

The very bright star near the centre is an exploding supernova.

## Questions

1 Describe the reaction that is the source of the Sun's energy.

2 Many people think that the Sun is a gigantic fire. Explain what is wrong with the statement 'The Sun burns hydrogen'.

3 Which is denser: a red giant or a white dwarf? Explain how you can tell.

## Key ideas

- *Gravitational forces cause parts of clouds of dust and gas in space to collapse and form new stars.*
- *Eventually, the Sun will expand to become a red giant, and then contract to become a white dwarf, before cooling to become a black dwarf.*

# The search for life

The Universe contains a vast number of stars, and many of these must have their own planets. Astronomers have already discovered dozens of planets in orbit around other stars. But is there any evidence that living creatures inhabit any of them? Is there life elsewhere in the Solar System?

## Looking for life

One way to look for life in the Solar System is to send a spacecraft. We can receive pictures back on Earth, or a robot can collect samples, analyse them and send the results by radio signals to Earth. Such spacecraft have visited some likely places, including Mars and one of Jupiter's moons called Europa.

Images and samples have been examined to see if they contain anything living, or any fossils. Atmospheric gases have been analysed to look for **oxygen**. So far, nothing has been found.

**a Explain why it is difficult to land a spacecraft on a planet that has no atmosphere.**

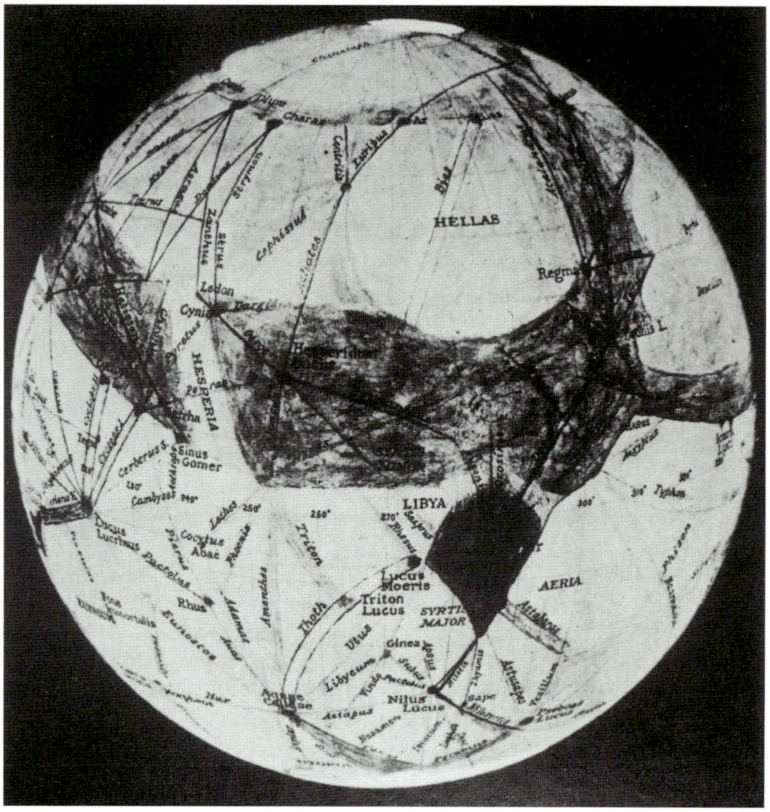


It was once thought that there are 'seas' and 'canals' on Mars

## Checking the atmosphere

When a spacecraft was sent to Jupiter, it looked back towards Earth. It could detect the oxygen in our atmosphere. Oxygen is a good indicator of life; on a dead planet, the oxygen has all been used up and is trapped in rocks. On Earth, plants produce the oxygen that animals need to live.

Scientists look for such signs of life in the atmospheres of other planets. They might also place a sample of material from another planet in a sealed container to see if the air around it changes because of the presence of **living organisms**.

**b Why are plants necessary if animals are to live?**

**Did you know?**

Venus has a very dense, acidic atmosphere. Robots that have been sent to land on Venus have not survived long enough to send back any useful information.

This spacecraft has landed on Mars. By checking soil and atmospheric gases, such a craft looks for signs of life.

## Listening for life

Because we broadcast radio and television signals around the world, we are sending messages out into space all the time. Observers on a different planet might detect these and wonder whether there was intelligent life on Earth.

Scientists have been listening out for radio signals coming from space for over 40 years. This is known as the Search for Extra-Terrestrial Intelligence, **SETI**. They hope to find meaningful, regular signals that would show that another civilisation is trying to make contact; so far, they have detected nothing but random 'noise'.

**Radio telescopes provide lots of useful information about stars and other galaxies, but so far they have not detected signals from intelligent creatures elsewhere in space.**

- Some scientists argue that there are billions of planets in the Universe. Many are similar to Earth, and so are likely to support life.
- Other scientists argue that, if there is intelligent life elsewhere, it should have made contact with us by now.

**c What do you think?**

## Questions

1 What do the letters 'SETI' stand for?

2 When the Earth formed, its atmosphere was mostly nitrogen, carbon dioxide and sulphur dioxide. Now it contains 20% oxygen. What has caused its composition to change?

3 Saturn has a rocky moon called Titan. List some ways in which we could look for evidence of life on Titan.

4 We could try to make contact with an alien civilisation by sending signals out into space. Do you think we should try this? What signals should we send?

### Key ideas

- *There are three ways of detecting other life forms.*
  - *Analysing the atmospheres of planets to see if they contain oxygen.*
  - *Using robots to carry out soil experiments.*
  - *Listening for radio signals.*

# 6:11 The Universe – its past and future

Measurements show that most of the other galaxies in the Universe are moving away from the Milky Way. The further away a galaxy is from us, the faster it is receding.

This leads to the conclusion that the Universe is expanding. Scientists use this to try to explain how the Universe arrived at its current state and what will happen to it in the future.

Nothing is certain – all we have are theories.

These galaxies are all moving away from each other.

## Steady state

In the **steady state theory** the Universe has properties that do not change. Its temperature and density stay the same. The theory accepts that the size of the Universe is changing, because it is known to be expanding. However, the expansion is described as being at a constant rate.

This expansion would cause the density of the Universe to gradually become less, so it would not be in a steady state. To make up for this, the theory states that new matter is being created all the time so that the density can stay the same.

However, the steady state theory does not quite fit all the evidence, so a new model is needed.

**a Which of these properties of the Universe are 'steady' according to the steady state theory?**

**density    mass    temperature    volume**

**Did you know?**
The Andromeda galaxy is the only galaxy that is known to be getting closer to the Milky Way.

## Big bang

If the galaxies today are moving apart, at one time they must have been much closer together. Most astronomers believe that it must have begun with everything squashed into a tiny space. Between 12 and 18 billion years ago, there was a giant explosion, the **big bang**.

All the matter in the Universe was created in the big bang. The expansion of the Universe is a result of the explosion, just as debris is flung out from an exploding firework.

Unlike the steady state theory, the big bang theory predicts that the Universe is becoming less dense and cooling as it expands.

**b What predictions about the future of the Universe are different in the steady state and big bang theories?**

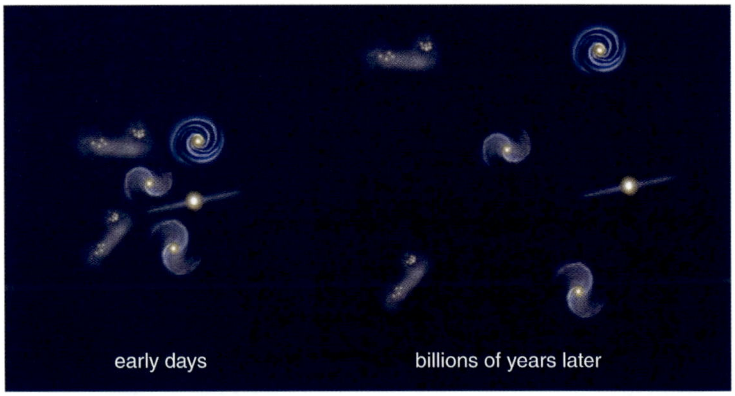


As time goes by, the Universe expands, so that the galaxies move farther and farther apart.

## The future

Just as we are uncertain about the history of the Universe, we cannot be sure about its future. There are three options.

- The Universe will carry on expanding and cooling.
- The Universe will reach a size where the expansion stops and stay at that size.
- The Universe will stop expanding and contract.

Which one will happen depends on how much mass the Universe contains – an unknown quantity. The greater the mass, the more likely it is that gravitational forces will stop the expansion and could cause the Universe to contract. At the moment, the evidence is that the Universe will carry on expanding forever.

**c Some astronomers think that the Universe could end in a 'big crunch'. What could happen to cause this?**

## Questions

**1 Which of these statements is about the steady state theory and which is about the big bang theory?**

**A The temperature of the Universe is constant.**

**B The temperature of the Universe is decreasing.**

**C The Universe must once have been very hot.**

**D New matter is always being made.**

**2 The lines on the graph show three possibilities for the size of the Universe.**

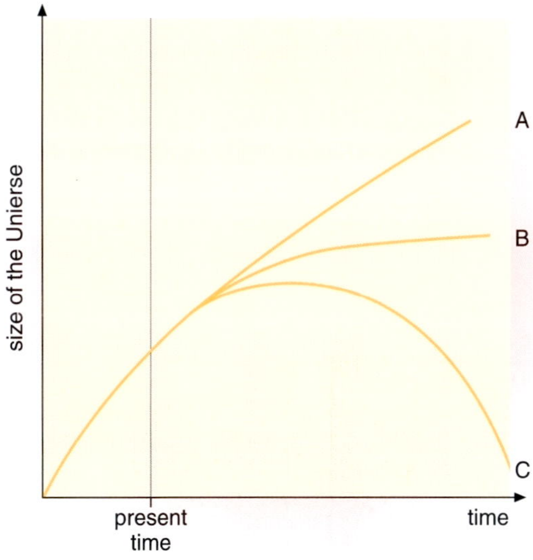


**a Describe what each line shows could happen in the future.**

**b Explain how the future of the Universe depends on the amount of mass that it contains.**

**3 How does the 'big bang' theory explain the movement of the galaxies?**

### Key ideas

- *The steady state theory of the Universe describes it as being in a largely unchanging state.*
- *The big bang theory describes how the Universe started in an enormous explosion and has been expanding and cooling ever since.*
- *The future of the Universe depends on the amount of mass that it contains.*

# 6:12 Atoms, elements, isotopes

If you could look very closely at an atom, how could you tell which element it was? Simply count the number of **protons** in the nucleus. Every atom of the same element has the same number of protons.

- An atom with 1 proton in its nucleus is an atom of hydrogen.
- An atom with 2 protons in its nucleus is an atom of helium.
- An atom with 6 protons in its nucleus is an atom of carbon.
- An atom with 92 protons in its nucleus is an atom of uranium.

If you look at the Periodic Table, you will see that hydrogen is element number 1, helium is number 2, and so on.

The number of protons in the nucleus is the **atomic number** of the atom.

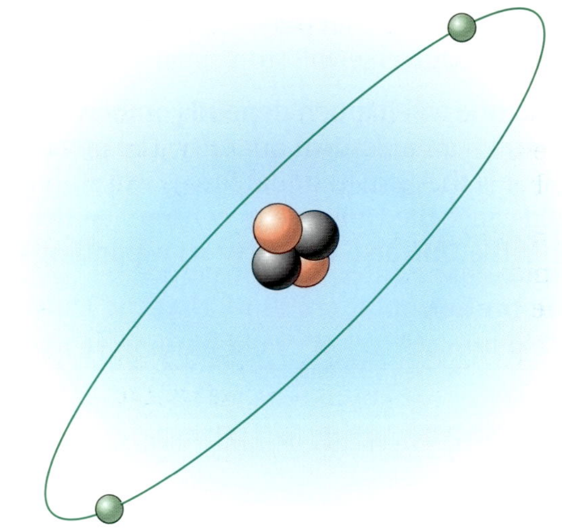

This atom has two protons in its nucleus; it is an atom of helium. (It also has two neutrons.)

**a A neutral atom has six electrons orbiting its nucleus. How many protons are in its nucleus? Which element is it an atom of? (Look in the list above.)**

## Isotopes

All carbon atoms have six protons in the nucleus. However, they are not all identical. Some have six **neutrons**, some have seven, and some have eight. The illustration shows these atoms. They are known as **isotopes** of carbon.

All these isotopes of carbon react with oxygen to form carbon monoxide and carbon dioxide. The only way that they are different is that the atoms have different masses.

The isotopes of an element all have:

- the same number of protons;
- different numbers of neutrons.

You can work out the **mass number** (**nucleon number**) of an isotope by adding up the total number of protons and neutrons (nucleons) in its nucleus.

**Did you know?**
Neutrons were discovered by James Chadwick in 1932. However, in 1920, Lord Rutherford had predicted their existence in order to explain the mass of the helium nucleus.

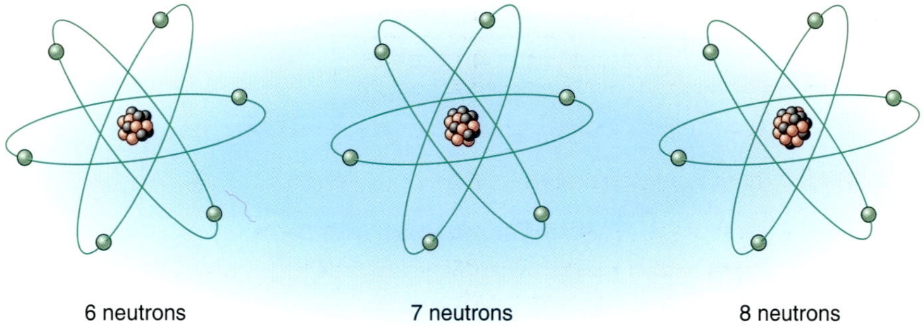


Atoms of three isotopes of carbon.

Mass (nucleon) number = number of protons + number of neutrons

**b An atom of oxygen has 8 protons and 8 neutrons in its nucleus. What is its mass number?**

**c Another atom has 8 protons and 9 neutrons in its nucleus. Is it an isotope of oxygen, or a different element?**

## Using symbols

Chemical symbols are a form of shorthand for scientists. They give quick ways of writing out chemical names and they are also used to show the number of atoms of each element in a particular molecule.

By adding numbers to the symbol, the structure of the nucleus can be shown.

The isotope of carbon that has 6 protons and 8 neutrons in the nucleus is often referred to as carbon-14, because its mass number is 14. In symbols it is written as:

number of nucleons

$$^{14}_{6}C$$

number of protons

The numbers on the left of the chemical symbol show the structure of the nucleus.

- The upper number is the mass number.
- The lower number is the atomic number.

**d How can you work out the number of neutrons from the symbol?**

## Questions

**1 The atoms of two isotopes of helium contain the following particles.**

| | | | |
|---|---|---|---|
| **Isotope 1** | 2 protons | 2 neutrons | 2 electrons |
| **Isotope 2** | 2 protons | 1 neutron | 2 electrons |

**a Draw diagrams to represent these two atoms.**

**b Explain why they are both atoms of the same element.**

**2 The table shows the particles in the nuclei of three different atoms.**

| | | |
|---|---|---|
| **Atom 1** | 7 protons | 7 neutrons |
| **Atom 2** | 6 protons | 7 neutrons |
| **Atom 3** | 7 protons | 6 neutrons |

**a Which two are isotopes of the same element?**

**b Which two have the same nucleon number?**

**3 Chlorine, chemical symbol Cl, has two common isotopes. They each have 17 protons in the nucleus. One has 18 neutrons and the other has 20.**

**Write down the symbols for these isotopes.**

### Key ideas

- *All atoms of an element have the same atomic number, the number of protons.*
- *Different isotopes of an element have different mass (nucleon) numbers.*
- *The atomic and mass numbers of an isotope are shown by numbers written on the left of the chemical symbol.*

# 6:13 Alpha, beta, gamma

There are radioactive substances all around us, and they give out radiation all the time. We cannot see the radiation that they give out. That is why it took scientists a long time to discover radioactivity.

The first person to notice the radiation coming from a radioactive substance was a French scientist called Henri Becquerel in 1896. He was studying rocks that contained uranium. Some of these rocks glow in the dark. He noticed that they made a photographic film go dark, even when the film was wrapped in black paper to stop light getting at it. Some kind of radiation was getting through the paper.

**a People are often very nervous about radioactive substances. Suggest why this is.**

## Absorbing radiation

Radioactive substances emit three main types of radiation.

- **Alpha (α) radiation** is stopped by a thin sheet of paper or a few centimetres of air.
- **Beta (β) radiation** passes easily through air or paper, but is stopped by a few millimetres of metal.
- **Gamma (γ) radiation** cannot be totally absorbed. It needs many centimetres of lead, or metres of concrete, to absorb most of it.

The diagram summarises this information.

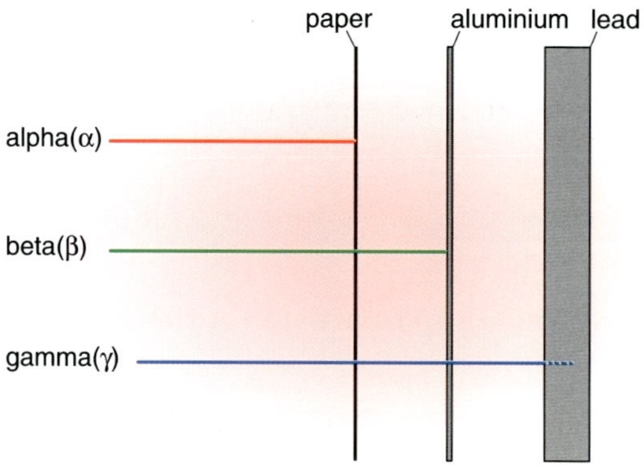


**b Which type of radiation is the most penetrating? Which is the least penetrating?**

## Monitoring thickness

Many factories make use of radiation to check the thickness of materials they are producing. For example, beta radiation can be used to check the thickness of plastic sheeting.

- If the sheeting is too thick, less beta radiation passes through it. The machine goes faster to make the sheeting thinner.
- If the sheeting is too thin, more beta radiation gets through to the detector. This tells the machine to go slower.

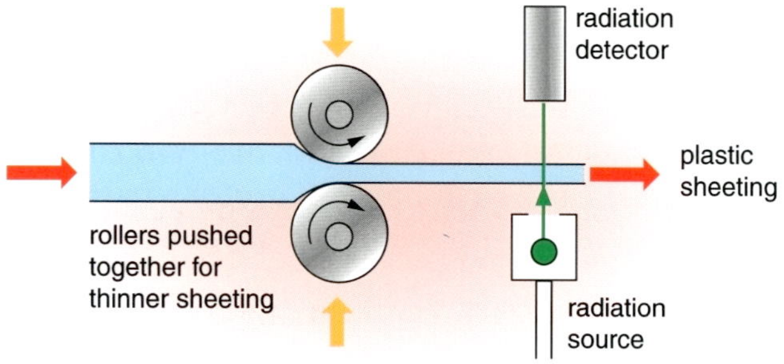


**c The thickness of metal sheeting can be monitored in the same way. Explain why gamma radiation must be used instead of beta radiation.**

## Radiation properties

Radioactivity is given out when an unstable nucleus changes to a more stable one. This decay is random – the decay of any particular nucleus cannot be predicted.

The three main types of radioactive emission have very different properties in terms of their mass and charge. This affects their ability to cause **ionisation**.

Alpha radiation is the largest and most massive type. It has a very short range in air because of the frequent collisions with air particles. When it collides with an air particle it loses energy and knocks one or more electrons off the air particle, creating positive and negative **ions**. This is shown in the diagram.

The properties of alpha, beta and gamma radiation are shown in the table.

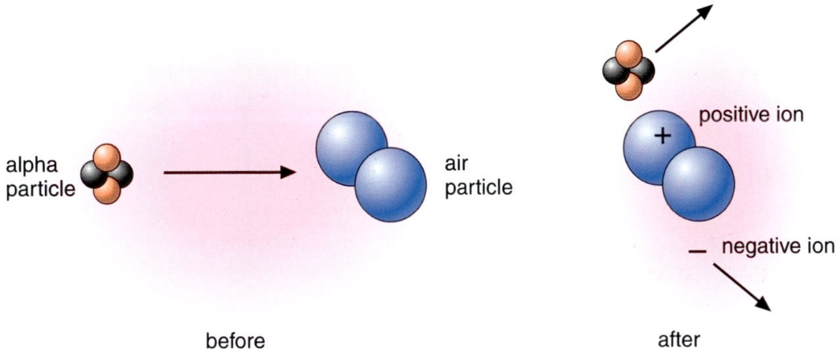


### Did you know?

The luminous dials on clocks and watches used to be painted using radioactive paint.

Years later, the factory workers developed throat cancers. They had been in the habit of licking the brushes to get a sharp point.

| Radioactive emission | Nature | Charge | Mass | Ability to cause ionisation |
|---|---|---|---|---|
| alpha particle | two neutrons and two protons | +2 | 4 × the mass of a proton | strong |
| beta particle | fast-moving electron | –1 | 1/2000 the mass of a proton | medium |
| gamma ray | short-wavelength electromagnetic radiation | 0 | 0 | weak |

**d Which emission has the greatest mass?**

## Questions

**1** Copy the table.

| | | |
|---|---|---|
| _____ easily absorbed | alpha radiation | $\alpha$ |
| | beta radiation | |
| _____ penetrating | gamma radiation | |

**a** In the spaces in the first column, write '**most**' or '**least**'.

**b** Complete the last column by writing the correct symbols for beta and gamma radiation.

**2** Which types of radiation are being described here? Choose from: **alpha**, **beta** and **gamma**.

**a** It is absorbed by a few millimetres of aluminium.

**b** It passes easily through air and through a centimetre of metal.

**c** It is absorbed by a thin sheet of paper.

**3** Explain why alpha radiation is the most harmful emission when inside the body and the least harmful when outside the body.

### Key ideas

- *Radioactive substances emit three main types of radiation: alpha, beta and gamma.*
- *Gamma radiation is the most penetrating.*
- *Alpha radiation is the most intensely ionising.*

# 6:14 Radiation all around us

There are radioactive substances all around us. This means that we are exposed to a low level of radiation all the time. This is called **background radiation**. The diagram shows the sources of background radiation.

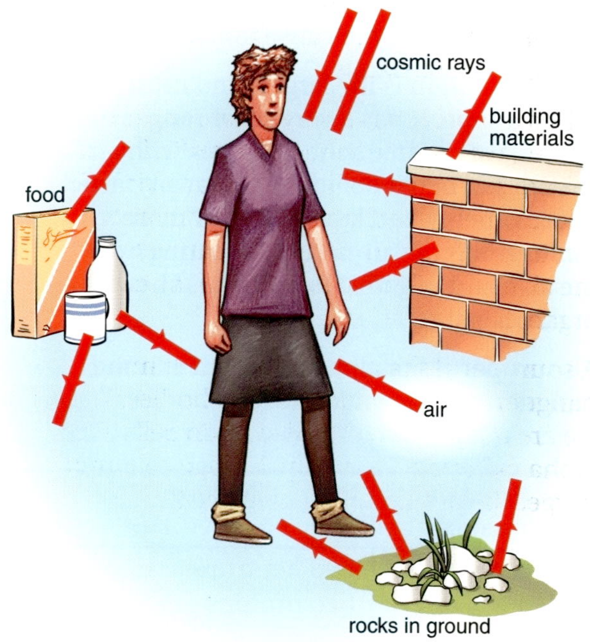


**Cosmic rays** come from sources far out in space. The atmosphere absorbs most of the cosmic radiation, but it is a serious problem for astronauts.

The food we eat contains small amounts of radioactive carbon-14. Plants absorb this radioactive carbon during photosynthesis.

**a Make a list of the sources of background radiation.**

## Radioactive gas

The air around us contains radon and other radioactive gases. In some parts of the country, people have fans under their floors to blow out the radon as it seeps up from underground granite. Radioactive rock is the main reason for the variation in background radiation shown in the diagram.

**b Suggest why the level of background radiation is high in Wales but low in the south east of England.**

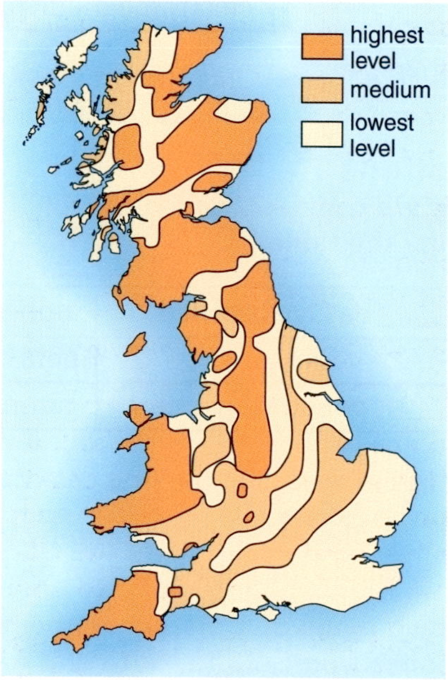


**The variation in background radiation in England, Scotland and Wales.**

## Cell damage and mutations

It is right to be cautious about radiation. It can cause cancer. The more radiation you are exposed to, the greater the chance that you will get cancer. This is what happens.

- Radiation is energetic and fast moving. When radiation smashes into neutral (uncharged) atoms or molecules, they may be left in a charged state – they are **ionised** (they have become ions).

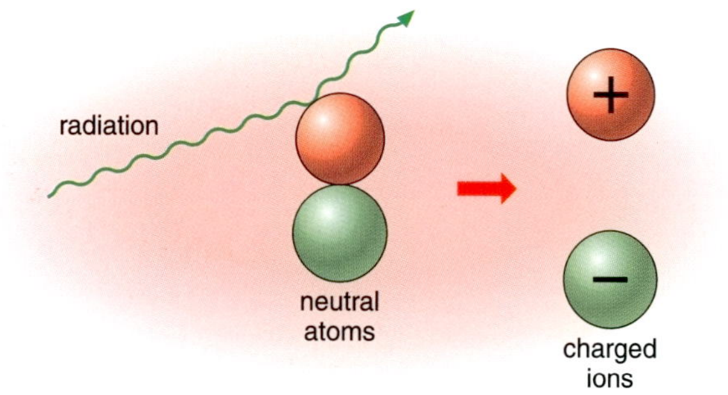


- If radiation damages a DNA molecule in a living cell, the cell may go out of control and start to divide over and over again. This is how a tumour (a cancer) develops.
- DNA is the chemical in the nucleus of every cell that controls the cell's actions and carries genetic information. If a gene is changed a **mutation** has occurred. This poses a particular threat if it happens in a sex cell as it could affect a future fetus.

**Did you know?**

We are all radioactive because we eat radioactive food. After death, the level of radioactivity decreases.

**c What is the difference between a neutral atom and an ion?**

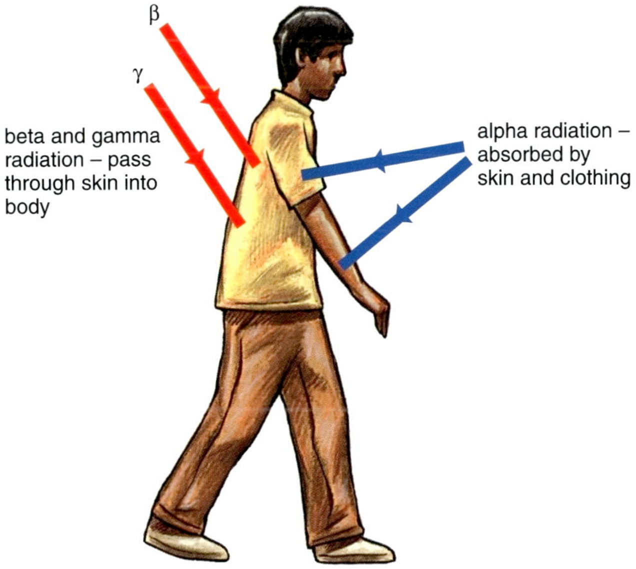


## Radiation protection

Because we understand radiation, we can find ways to protect ourselves from its effects. We can understand which are the most dangerous radioactive substances.

When a source of radiation is *outside* our bodies, we are safe from alpha radiation. It cannot penetrate the layer of dead skin cells that covers us. Beta and gamma radiation are more dangerous because they can penetrate the skin. Then they may damage the cells of organs within our bodies.

A source of alpha radiation is much more dangerous when it is *inside* our bodies. Then we are not protected by dead skin cells. The alpha radiation can easily damage cells and trigger the development of cancer.

**d Our skin is covered with a layer of dead cells. Which type of radiation does this protect us from? Why does it not protect us from the other two types?**

## Questions

**1** Copy these sentences and use words from the list to fill the gaps.

| | | |
|---|---|---|
| cancer | ions | background |
| mutation | neutral | |

We are constantly exposed to ________ radiation.

Radiation can change ________ atoms and molecules, so that they become charged ________.

If a molecule within a living cell is damaged, this may cause ________ or a ________.

**2** We are exposed to background radiation from many different sources. The pie chart shows the contributions that different sources make to background radiation in the UK. Put them in order, from biggest contributor to smallest.

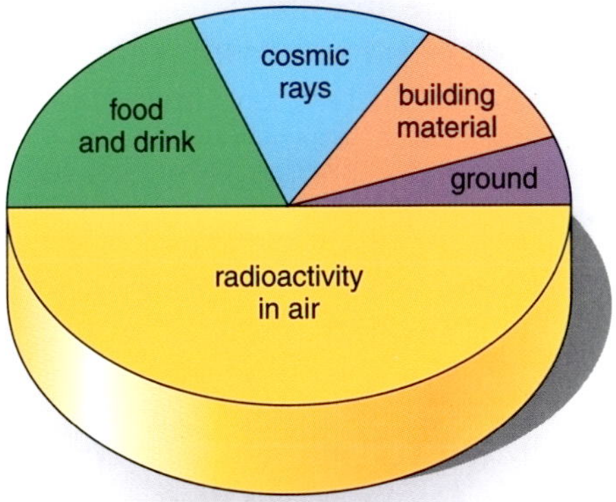


**3** Explain why a source of alpha radiation is dangerous when it is inside us, but not dangerous when it is outside us.

**4** The air contains a radioactive gas called radon. Radon is a source of alpha radiation. In some parts of the country, there is a lot of radon in the air. Explain why people in these areas are more likely to develop lung cancer.

### Key ideas

- *We are always exposed to background radiation from radioactive substances around us.*
- *Radiation can damage cells and cause cancer and mutations.*

# 6:15 Living with radiation

Syringes are sterilised using gamma radiation. Doctors need to be sure that the syringes they use are not contaminated with micro-organisms that might harm their patients. When syringes, in their plastic wrapper, are exposed to strong radiation, any micro-organisms are killed.

**a Explain why alpha radiation would not be suitable for sterilising syringes. Why is gamma radiation a good choice?**

Radiation may cause cancer, but it can also be used to cure it. When a patient has cancer, their tumour may be treated with a high dose of radiation to kill the cancer cells (see page 161).

## Using radioactivity at home

Most homes are fitted with **smoke alarms**. These are not affected by a small amount of smoke from cooking, but they sound a loud siren if there is smoke from a fire.

Some smoke detectors use radioactivity to detect smoke. The source used is an isotope of americium. This is an artificial source of radiation, it does not occur in nature. As the americium decays it emits alpha particles only.

The diagram shows the detection chamber and the circuit that is used.

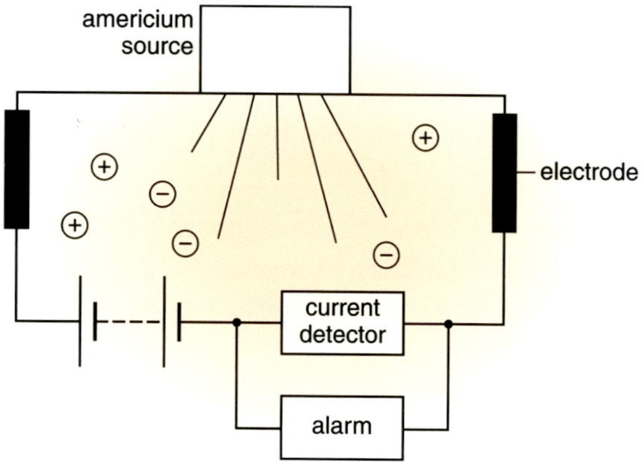


- The alpha particles ionise the air in the chamber and a current passes between the electrodes.
- When smoke enters the chamber the smoke particles stick to the air ions.
- This causes the ionisation current to fall.
- The fall in current is detected and this triggers the alarm.

**b Which way do the positive ions move in the detection chamber shown in the diagram?**

The americium source does not pose any hazard to people who pass near it. The alpha particles cannot penetrate the walls of the detection chamber.

**c Explain why it is important that the radioactive source used in a smoke alarm does not emit beta or gamma radiation.**

**Did you know?**

The most common cause of domestic fires is electrical faults due to faulty do-it-yourself wiring.

Always employ a qualified electrician to work on house wiring.

## Disposing of the waste

The use of radioactive materials in medicine, industry and nuclear power stations produces waste materials that are themselves radioactive. This cannot be burned or be placed in a landfill site like normal rubbish. Some of this waste will remain hazardous for thousands of years.

These drums contain intermediate level waste.

There are three categories of radioactive waste.

- **Low-level waste** consists of items such as laboratory clothing and packing materials. It is sealed in concrete containers that are stored in a concrete-lined room underground.
- **Medium-level waste** consists of radioactive sludge and contaminated equipment. It is surrounded by cement in stainless steel drums. At the moment it is kept in a temporary store until a suitable long-term disposal site is found.
- **High-level waste** consists of the remains from re-processed fuel. It is very radioactive. Because of this it generates a lot of heat. After being sealed in glass, it is placed in temporary storage. A permanent storage site is yet to be found. It needs to be buried deep underground where there is no risk that the site could be disturbed by geological movements of the earth. Such movement could break up the glass and result in radioactivity entering the water supply.

**d** Explain why the disposal of medium-level and high-level radioactive waste causes an environmental problem.

## Questions

**1 a** Give an example of an item that may be sterilised using gamma radiation.

**b** Explain what the word 'sterilised' means here.

**2 a** What type of radiation is given out by the americium source in a smoke alarm?

**b** Explain why this emission is not a hazard to people.

**3 a** In a smoke alarm, what happens to the ionisation current when smoke enters the detection chamber?

**b** Why does this happen?

**4 a** What does low-level radioactive waste consist of?

**b** Why is disposal of low-level waste much easier than the disposal of medium-level and high-level waste?

### Key ideas

- *Radioactivity is used for sterilising medical instruments.*
- *Smoke alarms use radioactivity to sound a siren when smoke is detected.*
- *There are problems in disposing of radioactive waste safely.*

# End of module questions

## Low demand

1. Which of the following is an electromagnetic wave?
   **A** light
   **B** sound
   **C** infrasound
   **D** ultrasound

2. When travelling in a vacuum, all electromagnetic waves have the same
   **A** amplitude
   **B** frequency
   **C** speed
   **D** wavelength

3. In a longitudinal wave the vibrations are
   **A** along the direction of travel
   **B** at right angles to the direction of travel
   **C** at 45° to the direction of travel
   **D** at 60° to the direction of travel

4. Which graph represents a digital signal?

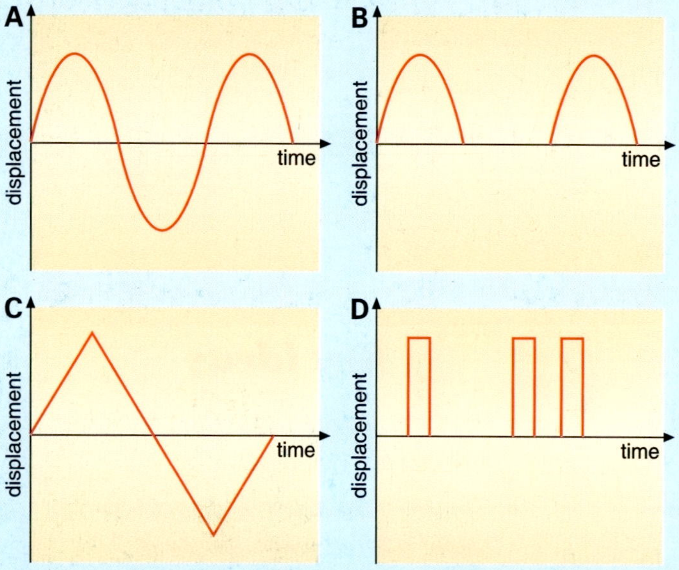


5. Infra-red radiation is used to
   **A** cook food
   **B** detect forged bank notes
   **C** detect a fracture in a bone
   **D** treat cancer

The diagram represents the electromagnetic spectrum.

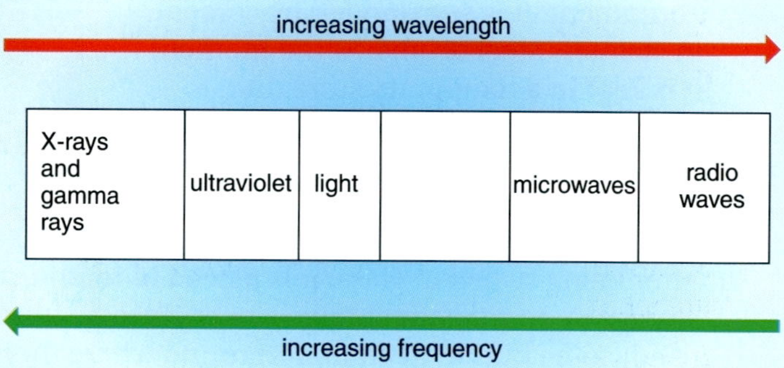


6. Which type of wave fits into the space between light and microwaves?
   **A** infrared
   **B** infraviolet
   **C** ultrared
   **D** television

7. As the wavelength of the waves increases, their frequency
   **A** decreases
   **B** doubles
   **C** halves
   **D** stays the same

8. Which type of wave is used by a television remote control?
   **A** ultraviolet
   **B** light
   **C** infrared
   **D** radio

**9** Which of these travels as a longitudinal wave?

**A** gamma rays

**B** light

**C** sound

**D** X-rays

**10** Sound that has a frequency of 25 000 Hz is an example of

**A** loud sound

**B** quiet sound

**C** infrasound

**D** ultrasound

**11** A galaxy is a collection of

**A** meteors

**B** moons

**C** planets

**D** stars

**12** Radioactivity is due to the breakdown of an unstable

**A** electron

**B** neutron

**C** nucleus

**D** proton

## Standard demand

**13** A signal that can only have the values 0 or 1 is

**A** an analogue signal

**B** an amplitude signal

**C** a digital signal

**D** a frequency signal

**14** Which electromagnetic wave has the shortest wavelength?

**A** gamma radiation

**B** infrared radiation

**C** light

**D** radio waves

**15** The X-rays used to photograph a person can pass through

**A** bone but not flesh

**B** flesh but not bone

**C** both flesh and bone

**D** neither flesh nor bone

**16** Which option gives the colours in order of increasing wavelength?

**A** red green blue

**B** red blue green

**C** blue red green

**D** blue green red

**17** The diagram represents a wave on a rope.

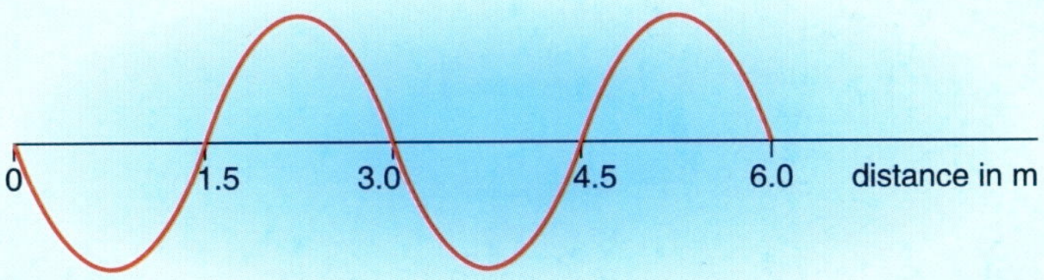


The wavelength is

**A** 1.5 m

**B** 3.0 m

**C** 4.5 m

**D** 6.0 m

**18** Which type of electromagnetic wave is used to show the ink markings from a security pen?

**A** gamma radiation

**B** ultraviolet radiation

**C** light

**D** infrared radiation

**19** Which diagram shows how light passes through a sheet of glass?

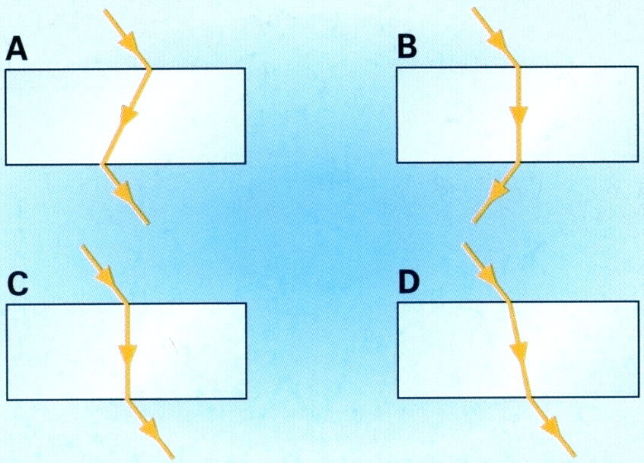

**20** Ultrasound is a compression wave that

**A** travels faster than normal sound

**B** travels slower than normal sound

**C** has a higher frequency than normal sound

**D** has a lower frequency than normal sound

**21** The force that keeps a planet in orbit around the Sun is

**A** electric

**B** friction

**C** gravitational

**D** magnetic

**22** Io is a natural satellite of the planet Jupiter.

Io is

**A** a galaxy

**B** a moon

**C** a planet

**D** a star

**23** After its main sequence, our Sun is likely to become

**A** a black giant

**B** a blue giant

**C** a red giant

**D** a white giant

**24** An atom of $^{17}_{8}O$ has

**A** 8 neutrons in the nucleus

**B** 8 neutrons in orbit around the nucleus

**C** 8 protons in the nucleus

**D** 8 protons in orbit around the nucleus

## The periodic table of elements

**KEY**

| | |
|---|---|
| Mass number A | 1 |
| Symbol | **H** |
| Name | Hydrogen |
| Atomic number (Proton number) Z | 1 |

| 1 | 2 | | | | | | | | | | | 3 | 4 | 5 | 6 | 7 | 0 |
|---|---|---|---|---|---|---|---|---|---|---|---|---|---|---|---|---|---|
| | | | | | | | | | | | | | | | | | 4<br>**He**<br>Helium<br>2 |
| 7<br>**Li**<br>Lithium<br>3 | 9<br>**Be**<br>Beryllium<br>4 | | | | | | | | | | | 11<br>**B**<br>Boron<br>5 | 12<br>**C**<br>Carbon<br>6 | 14<br>**N**<br>Nitrogen<br>7 | 16<br>**O**<br>Oxygen<br>8 | 19<br>**F**<br>Fluorine<br>9 | 20<br>**Ne**<br>Neon<br>10 |
| 23<br>**Na**<br>Sodium<br>11 | 24<br>**Mg**<br>Magnesium<br>12 | | | | | | | | | | | 27<br>**Al**<br>Aluminium<br>13 | 28<br>**Si**<br>Silicon<br>14 | 31<br>**P**<br>Phosphorus<br>15 | 32<br>**S**<br>Sulphur<br>16 | 35<br>**Cl**<br>Chlorine<br>17 | 40<br>**Ar**<br>Argon<br>18 |
| 39<br>**K**<br>Potassium<br>19 | 40<br>**Ca**<br>Calcium<br>20 | 45<br>**Sc**<br>Scandium<br>21 | 48<br>**Ti**<br>Titanium<br>22 | 51<br>**V**<br>Vanadium<br>23 | 52<br>**Cr**<br>Chromium<br>24 | 55<br>**Mn**<br>Manganese<br>25 | 56<br>**Fe**<br>Iron<br>26 | 59<br>**Co**<br>Cobalt<br>27 | 59<br>**Ni**<br>Nickel<br>28 | 63<br>**Cu**<br>Copper<br>29 | 64<br>**Zn**<br>Zinc<br>30 | 70<br>**Ga**<br>Gallium<br>31 | 73<br>**Ge**<br>Germanium<br>32 | 75<br>**As**<br>Arsenic<br>33 | 79<br>**Se**<br>Selenium<br>34 | 80<br>**Br**<br>Bromine<br>35 | 84<br>**Kr**<br>Krypton<br>36 |
| 85<br>**Rb**<br>Rubidium<br>37 | 88<br>**Sr**<br>Strontium<br>38 | 89<br>**Y**<br>Yttrium<br>39 | 91<br>**Zr**<br>Zirconium<br>40 | 93<br>**Nb**<br>Niobium<br>41 | 96<br>**Mo**<br>Molybdenum<br>42 | 99<br>**Tc**<br>Technetium<br>43 | 101<br>**Ru**<br>Ruthenium<br>44 | 103<br>**Rh**<br>Rhodium<br>45 | 106<br>**Pd**<br>Palladium<br>46 | 108<br>**Ag**<br>Silver<br>47 | 112<br>**Cd**<br>Cadmium<br>48 | 115<br>**In**<br>Indium<br>49 | 119<br>**Sn**<br>Tin<br>50 | 122<br>**Sb**<br>Antimony<br>51 | 128<br>**Te**<br>Tellurium<br>52 | 127<br>**I**<br>Iodine<br>53 | 131<br>**Xe**<br>Xenon<br>54 |
| 133<br>**Cs**<br>Caesium<br>55 | 137<br>**Ba**<br>Barium<br>56 | 139<br>**La**<br>Lanthanum<br>57 | 178<br>**Hf**<br>Hafnium<br>72 | 181<br>**Ta**<br>Tantalum<br>73 | 184<br>**W**<br>Tungsten<br>74 | 186<br>**Re**<br>Rhenium<br>75 | 190<br>**Os**<br>Osmium<br>76 | 192<br>**Ir**<br>Iridium<br>77 | 195<br>**Pt**<br>Platinum<br>78 | 197<br>**Au**<br>Gold<br>79 | 202<br>**Hg**<br>Mercury<br>80 | 204<br>**Tl**<br>Thallium<br>81 | 207<br>**Pb**<br>Lead<br>82 | 209<br>**Bi**<br>Bismuth<br>83 | 210<br>**Po**<br>Polonium<br>84 | 210<br>**At**<br>Astatine<br>85 | 222<br>**Rn**<br>Radon<br>86 |
| 223<br>**Fr**<br>Francium<br>87 | 226<br>**Ra**<br>Radium<br>88 | 227<br>**Ac**<br>Actinium<br>89 | | | | | | | | | | | | | | | |

Elements 58–71 and 90–103 have been omitted.

The value for mass number is normally that of the commonest isotope, e.g. $^{35}Cl$ not $^{37}Cl$.

Bromine is approximately equal proportions of $^{79}Br$ and $^{81}Br$.

# Key words

The key words listed below are terms that you need to understand for your exams. Copying them out and providing definitions from the relevant spread may help with your revision.

## Module 1 – The human body: action and control

**Spread 1:1**
carbohydrase
carbohydrates
digestive system
enzymes
fats
lipase
protease
protein

**Spread 1:2**
bile
emulsification
emulsion

**Spread 1:3**
absorption
faeces
fibre
plasma
villus (plural villi)

**Spread 1:4**
Plasma
platelets
red blood cells
white blood cells

**Spread 1:5**
antibiotic
bacteria
chromosome
gene
microbe
virus

**Spread 1:6**
cholera
toxins

**Spread 1:7**
antibodies
antigens
antitoxins
cilia
clot
lymphocytes
mucus
phagocytes
toxins

**Spread 1:8**
brain
central nervous system
coordinator
environment
nerve
nervous system
neurone
receptors
sense organs
stimulus

**Spread 1:9**
co-ordinators
reflex action

**Spread 1:10**
cornea
iris
lens
optic nerve
pupil
retina

**Spread 1:11**
addiction
drug
hallucinogens
painkillers
sedatives
stimulants
withdrawal symptoms

**Spread 1:12**
bronchioles
carbon monoxide

**Spread 1:13**
homeostasis
hormone
thermoregulatory centre

**Spread 1:14**
bladder
dialysis
kidney
tubule

## Module 2 – Inheritance and survival

**Spread 2:1**
characteristic
chromosome
diploid number
inherited
species

**Spread 2:2**
gamete
haploid number
ovaries
ovum
sperm
testes

**Spread 2:3**
asexual
clone
fertilisation
sex chromosome
sexual
zygote

**Spread 2:4**
gland
menstrual cycle
menstruation
oestrogen
progesterone
sex hormone
testosterone

**Spread 2:5**
base
DNA
double helix
gene
replication
strand

**Spread 2:6**
chromosome
gene
gene therapy
Human Genome Project

**Spread 2:7**
allele
dominant
recessive

**Spread 2:8**
environment
inherited disorder
ionising radiation
mutation
skin cancer

**Spread 2:9**
genetically modified
GM crops
weed killer

**Spread 2:10**
adaptation
evaporation
feature
gills
scales
streamlined

**Spread 2:11**
competition
population
predator
prey
resources

**Spread 2:12**
evolution
evolved
extinct
natural selection

**Spread 2:13**
artificial insemination
selective breeding

**Spread 2:14**
acid rain
fossil fuels
pollution

## Module 3 – Chemical patterns

**Spread 3:1**
atom
molecule

**Spread 3:2**
atomic number
electrons
neutrons
nucleus
protons

**Spread 3:3**
electron shell
electronic structure

**Spread 3:4**
atomic mass
triad

**Spread 3:5**
group
period
Periodic Table

**Spread 3:6**
properties

**Spread 3:7**
diatomic
halogen

**Spread 3:8**
displacement reaction

**Spread 3:9**
reaction rate

**Spread 3:10**
concentration
surface area

**Spread 3:11**
collision

**Spread 3:12**
catalyst
denatured
enzyme

## Module 4 – Chemistry in action

**Spread 4:1**
Fossil fuel
Hydrocarbon

**Spread 4:2**
Fractions
Fractional distillation

**Spread 4:3**
Carbon monoxide
Complete combustion
Incomplete combustion

**Spread 4:4**
Bitumen
Cracking
Thermal decomposition

**Spread 4:5**
Polymerisation
Polymer

**Spread 4:6**
Biodegradable

**Spread 4:7**
Fermentation
Raising agent
Yeast

**Spread 4:8**
Carbohydrase
Lipase
Protease

**Spread 4:9**
Cement
Concrete
Quicklime
Raw material
Thermal decomposition

**Spread 4:10**
Fertiliser
Neutralisation
Salt

## Module 5 – Energy and electricity

**Spread 5:1**
ammeter
battery
current
parallel
resistance
series
voltage
voltmeter

**Spread 5:2**
variable resistor

**Spread 5:3**
filament
proportion
resistor

**Spread 5:4**
diode
light-dependent resistor
thermistor

**Spread 5:5**
cable grip
conductor
double-insulated
earth
fuse
insulator
live
neutral
plug

**Spread 5:6**
alternating current (a:c:)
direct current (d:c:)
earth wire
fuse

**Spread 5:7**
kilowatt (kW)
kilowatt-hour (kW h)
power
unit

**Spread 5:8**
conduction
convection current
electromagnetic radiation
infrared radiation
immersion heater
radiant energy
thermostat

**Spread 5:9**
dynamo
electromagnet
electromagnetic induction
generator
magnetic field

**Spread 5:10**
power line
pylon
step-up transformer
step-down transformer
transformer

**Spread 5:11**
acid rain
boiler
fossil fuel
greenhouse effect
non-renewable energy

**Spread 5:12**
hydroelectric power
kinetic energy
renewable energy source
tidal power
wind farm
wave power
wind turbine

**Spread 5:13**
plutonium
radiation
radioactive
radioactive waste
reactor
uranium

**Spread 5:14**
solar cell
solar heating

**Spread 5:15**
efficient
energy-efficient
inefficient

**Spread 5:16**
cavity wall insulation
cost effective
double glazing
loft insulation
pay-back period

## Module 6 – Waves, atoms and space

**Spread 6:1**
amplitude
frequency
longitudinal
transverse
vibration
wavelength

**Spread 6:2**
electromagnetic spectrum
electromagnetic wave
gamma rays
infrared rays
light
microwaves
radio waves
spectrum
ultraviolet rays
X-rays

**Spread 6:3**
infrared radiation
microwaves

**Spread 6:4**
gamma rays
ultraviolet radiation
X-rays

**Spread 6:5**
analogue signal
digital signal
normal line
optical fibre
refraction

**Spread 6:6**
echo sounding
frequency range
ultrasound

**Spread 6:7**
ellipse
gravitational attraction
gravitational field
gravitational field strength
gravitational pull
Moon
orbit
planet
satellite
Solar System

**Spread 6:8**
comet
galaxy
Milky Way
star
Universe

**Spread 6:9**
black dwarf
main sequence
nebula
nuclear fusion
red giant
white dwarf

**Spread 6:10**
living organisms
oxygen
SETI

**Spread 6:11**
big bang
steady state theory

**Spread 6:12**
atomic number
isotope
mass number (nucleon number)
neutron
proton

**Spread 6:13**
alpha radiation
beta radiation
gamma radiation
ionisation
penetration

**Spread 6:14**
background radiation
cosmic rays
ionise
mutation

**Spread 6:15**
intermediate-level waste
high-level waste
low-level waste
smoke alarm
sterilise

# Index